建筑承包商标准信函格式(第三版)

[英]大卫·查贝尔 著
方志达 余培明 译
夏瑞康 周常明
方志达 校

中国建筑工业出版社

著作权合同登记图字：01-2003-5043 号

图书在版编目(CIP)数据

建筑承包商标准信函格式/(英)查贝尔著；方志达等译．
三版—北京：中国建筑工业出版社，2004
ISBN 7-112-06319-1

Ⅰ．建… Ⅱ．①查… ②方… Ⅲ．建筑工程—承包—信函—范文 Ⅳ．TU723.1

中国版本图书馆 CIP 数据核字（2004）第 005761 号

译自"David Chappell: Standard Letters for Building Contractors, 3rd edition"
ⓒ D. M. Chappell 1987, 1942, 2003
本书的翻译出版得到英国 Blackwell Publishing Ltd, Oxford 的大力支持。

责任编辑：丁洪良　董苏华
责任设计：刘向阳
责任校对：赵明霞

建筑承包商标准信函格式（第三版）
[英]大卫·查贝尔　著
方志达　余培明　夏瑞康　周常明　译
方志达　校
*
中国建筑工业出版社出版、发行（北京西郊百万庄）
新　华　书　店　经　销
北京海通创为图文设计有限公司制作
北京建筑工业印刷厂印刷
*
开本：787×1092 毫米　1/16　印张：22½　字数：543 千字
2004 年 6 月第一版　2004 年 6 月第一次印刷
定价：45.00 元
ISBN 7-112-06319-1
TU·5574(12333)

版权所有　翻印必究
如有印装质量问题，可寄本社退换
（邮政编码　100037）

本社网址：http：//www.china-abp.com.cn
网上书店：http：//www.china-building.com.cn

目 录

绪论

第一章 招投标 1

函件 1   关于要求列入邀请投标人名单,致建筑师   2
函件 2   关于要求列入邀请投标人名单,但未有回音,致建筑师   4
函件 3   同意投标,致建筑师   5
函件 4   如果承包商不愿投标,致建筑师   6
函件 5   如果在被列入邀请投标人名单之前要求承包商提供材料,致建筑师   7
函件 6   如果通知承包商投标日期推迟,承包商仍愿参与投标时,致建筑师   8
函件 7   如果承包商得知投标日期被推迟,且不愿意再参与投标时,致建筑师   9
函件 8   确认收到投标文件,致建筑师   10
函件 9   关于投标期间的质疑,致建筑师   11
函件 10   关于要求延长投标期限,致建筑师   12
函件 11   撤回投标资格,致建筑师   13
函件 12   致建筑师,根据方案 1 或 2 确认报价   14
函件 13   根据方案 1,要求撤回报价时,致建筑师   15
函件 14   根据方案 2,要求修改报价时,致建筑师   16
函件 15a   如果投标书被接受,致建筑师   17
函件 15b   如果投标书被接受,致建筑师   18
函件 16   如果其他承包商的投标书被接受时,致建筑师   19

第二章 合同文件 20

函件 17   寄还合同文件时,致建筑师   21
函件 18   如果合同文件中出现错误,致建筑师   22
函件 19   如果合同文件中出现错误,而先前的投标书未被接受,致建筑师   23
函件 20   如果要求承包商在合同文件签署之前开工,致建筑师   24
函件 21   如果合同尚未签订且付款证书签证期已到,致建筑师   25
函件 22   如果要求承包商签署担保书,致建筑师   27

| 函件 23 | 如果要求承包商提供履约保函,致建筑师 | 29 |

第三章 保险   31

函件 24a	有关承包商的保险,致建筑师	32
函件 24b	有关承包商保险,致建筑师	34
函件 24c	在接受标书或续保通知后 21 天内,致雇主	36
函件 24d	有关承包商保险,致雇主	37
函件 25	保险人被认可后,致建筑师	38
函件 26	有关雇主责任险,致建筑师	39
函件 27	承担雇主责任险的保险人被认可后,致建筑师	40
函件 28	如需因承包商工期延误而给雇主造成的损失投保,致建筑师	41
函件 29	当承包商误期损害赔偿金的保险报价被接受时,致建筑师	42
函件 30	当联合消防规范的补救措施作为变更事项,并要采取应急行动时,致建筑师	43
函件 31a	有关雇主保险,致雇主	44
函件 31b	有关雇主保险,致雇主	45
函件 31c	有关雇主保险,致雇主	46
函件 32a	当已投保的风险发生并造成损害时,致建筑师与雇主	47
函件 32b	当已投保的风险发生并造成损害时,致建筑师与雇主	48
函件 33a	如果雇主未能使保险继续有效,致雇主	49
函件 33b	如果雇主未能使保险继续有效,致雇主	51
函件 33c	如果雇主未能使保险继续有效,致雇主	53

第四章 施工现场活动   54

函件 34a	致建筑师,有关负责人或代理人事项	55
函件 34b	致建筑师,有关现场经理的任命	56
函件 35	致建筑师,有关批准现场经理的撤免或更换	57
函件 36	若要求提供现场人员的姓名及地址,致建筑师	58
函件 37	若需要出入证件,致建筑师	59
函件 38a	若在合同规定日或延期日未能提供工地现场占有权,致雇主	60
函件 38b	若在规定日期没有移交工地,致雇主	62
函件 38c	若得到通知在规定日期不能移交工地,而且延期条款不适用时,致雇主	64
函件 39	若工地占有日提前时,致建筑师	65
函件 40	致雇主,同意聘用其他人员	66
函件 41	致雇主,拒绝聘用其他人员	67
函件 42	有关现场会议纪要的内容,致建筑师	68

目 录

函件 43a	致建筑师,提交总进度计划	69
函件 43b	致建筑师,提交总进度计划	70
函件 44	致建筑师,提交总进度计划的修改稿	71
函件 45	致建筑师,提交承包商的结算单	72
函件 46	若指令要求实施附件中未列明的工程,致建筑师	74
函件 47	若建筑师的指令严重影响规定工程的实施,致建筑师	75
函件 48	致建筑师,请求提供资料	77
函件 49	若图纸上有关放线资料不全,致建筑师	78
函件 50	致建筑师,要求确认放线定位无误	79
函件 51	当迟收到资料时,致建筑师	80
函件 52	若按资料提交时间表未收到资料时,致建筑师	81
函件 53	当图纸上发现设计错误时,致建筑师	82
函件 54	若建筑师拒绝修改设计错误时,致建筑师	83
函件 55	若承包商提供图纸时,致建筑师	84
函件 56	当他退回承包商的图纸并提出意见时,致建筑师	85
函件 57	当承包商提交图纸听取意见时,致建筑师	86
函件 58	当建筑师退回承包商的图纸并提有意见时,致建筑师	87
函件 59	若建筑师退回承包商的图纸并提出不合理的意见时,致建筑师	88
函件 60	若建筑师不能在规定的时限内退回承包商的图纸时,致建筑师	89
函件 61	若发现文件之间有矛盾,致建筑师	90
函件 62	若发现"雇主要求"文件内有矛盾,致建筑师	91
函件 63	若"承包商方案"文件中有不一致之处,致建筑师	92
函件 64	若发现"雇主要求"和"承包商方案"文件之间有矛盾,致建筑师	93
函件 65	致建筑师,说明法律规定和其他文件之间的不一致	94
函件 66	致建筑师,说明法律规定与其他文件之间的矛盾	95
函件 67	当要求处理符合法律法规规定的突发事件时,致建筑师	96
函件 68	若基准日后法律规定有变化时,致建筑师	97
函件 69	若基准日后决定控制开发项目时,致建筑师	98
函件 70	致雇主,反对任命建筑师的替代人选	99
函件 71	若建筑师的替代人选尚未任命时,致雇主	100
函件 72	致雇主,反对工料测量师替代人选的任命	102
函件 73	致建筑师,关于工程管理人员在现场签发的指示	103
函件 74	致建筑师,关于工程管理人员在现场签发的指令	104
函件 75	致建筑师,关于工程管理人员毁坏工程或材料	105
函件 76	当若干"专家"工作人员要参观工地时,致建筑师	107
函件 77	当工程管理人员直接指挥施工人员时,致建筑师	109
函件 78	致工料测量师,提交价格报表	110

函件 79	当价格报表没有全部被接受时,致工料测量师	112
函件 80	致工料测量师,提交按第 13A 条的报价	114
函件 81	致建筑师,有关计日工作凭证的核实	116
函件 82	若对工程是否属于变更或包括在合同内产生不同意见时,致雇主	117
函件 83	致建筑师,要求说明授权签发指令的条款	118
函件 84	致建筑师,确认口头指令	119
函件 85	致建筑师,要求确认口头指令	120
函件 86	若口头指令未得到书面确认时,致建筑师	121
函件 87	致建筑师,反对改变责任或规定的指令	122
函件 88a	致建筑师,收到要求执行指令的 7 天通知	124
函件 88b	致建筑师,收到要求执行指令的通知	126
函件 89	致建筑师,运走未使用的材料	128
函件 90	若材料采购不到时,致建筑师	129
函件 91	当工程或材料或货物验收不合格时,致建筑师	130
函件 92	若承包商反对执行按第 3.13.1 条发出的指令时,致建筑师	132
函件 93	在隐蔽工程打开检查后,致建筑师	133
函件 94	当挖方工程可供检查时,致建筑师	134
函件 95	当指令拆除缺陷工程之后又签发了指令时,致建筑师	135
函件 96	当指令清除缺陷工程之后又签发要求打开隐蔽工程的指令时,致建筑师	137
函件 97	若为拆除缺陷工程签发指令,致建筑师	139
函件 98	当工程将被隐蔽时,致建筑师	140
函件 99	当工地现场发现文物时,致建筑师	141

第五章 工程款支付 143

函件 100a	致建筑师,附上期中付款申请	144
函件 100b	致建筑师,附上期中付款申请	145
函件 101	致工料测量师,提交评估申请	146
函件 102	若工料测量师未对估价申请做出回应时,致建筑师	148
函件 103	若期中付款证书未签发,致建筑师	150
函件 104	若期中付款证书签证金额不足,致建筑师	151
函件 105	若未全额付款,且未发出拒付通知,致雇主	152
函件 106	若未在规定期限内支付预付款,致雇主	153
函件 107	若评估未按标价的分项工程表进行时,致建筑师	154
函件 108	致建筑师,要求为场外材料付款	155
函件 109	致雇主,提前 7 天发出暂停施工的通知	157
函件 110	若无视暂停施工的通知,7 天内仍未全额付款时,致雇主	158

函件 111	致雇主,要求支付拖欠款项的利息	159
函件 112	致雇主,要求把保留金存入独立的银行账户之内	160
函件 113	若未将保留金存入独立的银行账户,致雇主	161
函件 114	致建筑师,附上为准备最终付款证书所需的所有资料	162
函件 115	致建筑师,附上最终账单	163
函件 116a	若未在预定日期签发最终付款证书,致建筑师	164
函件 116b	若未在预定日期签发最终付款证书,致建筑师	166
函件 116c	若未在预定日期签发最终付款证书,致建筑师	167

第六章　工期顺延　　168

函件 117	若出现工期延误,但又没有工期顺延的理由时,致建筑师	169
函件 118	若造成工期延误的事件已停止,但又没有理由顺延工期时,致建筑师	170
函件 119a	若出现工期延误,并有工期顺延的理由时,致建筑师	171
函件 119b	若出现工期延误,并有工期顺延的理由时,致建筑师	172
函件 120a	致建筑师,提供工期顺延所需的进一步详细材料	174
函件 120b	致建筑师,提供工期顺延所需的进一步详细材料	175
函件 121	当建筑师要求提供进一步详细材料以批准工期的顺延,致建筑师	176
函件 122	当建筑师不合理地要求提供进一步详细材料以批准工期的顺延,致建筑师	177
函件 123	若建筑师所批准的工期顺延时间不够,致建筑师	179
函件 124	若建筑师所批准的工期顺延时间不够,并且不打算重新考虑时,致建筑师	180
函件 125	若未在规定时间内批准工期顺延,致建筑师	182
函件 126	若建筑师拖延工期顺延的批准,致建筑师	184
函件 127a	若建筑师尚未审查工期顺延申请,致建筑师	186
函件 127b	若建筑师尚未审查工期顺延申请,致建筑师	188
函件 127c	若建筑师对工期顺延申请无最终决定时,致建筑师	189
函件 128	若建筑师认为承包商未尽力而为时,致建筑师	190
函件 129	若雇主错扣了工期延误损失赔偿费用,致雇主	191
函件 130	若退回损失赔款时未含利息,致雇主	192

第七章　损失和(或)费用　　193

函件 131a	致建筑师,申请获得损失和(或)费用的付款	194
函件 131b	致建筑师,根据补充条款,申请获得损失和(或)费用的付款	195
函件 131c	致建筑师,申请获得损失和(或)费用的付款	196
函件 131d	致建筑师,为费用申请补偿款	197

函件 132	致建筑师,提供有关损失和(或)费用的进一步细节	198
函件 133	致建筑师或工料测量师,附上损失和(或)费用的详细资料	200
函件 134	致工料测量师,提供用于计算费用的资料	201
函件 135a	若金额核定拖延,致建筑师	202
函件 135b	若金额核定拖延,致建筑师	203
函件 136	若核定金额太少,致建筑师	204
函件 137	关于依据普通法的索赔,致雇主	205
函件 138	关于依据普通法的索赔,致雇主	206

第八章　合同终止、仲裁、裁决和工程竣工　　207

函件 139	当发出违约通知时,致雇主或建筑师	208
函件 140	当发出合理的违约通知时,致雇主或建筑师	210
函件 141	如果已发出不成熟的合同终止通知函时,致雇主	211
函件 142	致雇主,合同终止前发出违约通知	212
函件 143	致雇主,在发出违约通知后终止雇佣关系	213
函件 144	致雇主,因雇主破产而终止雇佣关系	214
函件 145	当任何一方提出合同终止时,致雇主	215
函件 146	致雇主,已投保的风险带来损害后决定终止雇佣关系	216
函件 147	致雇主,发出有意将争端事项诉诸裁决的通知	217
函件 148	致指定机构,请求任命裁决人	218
函件 149	致裁决人,附上提交裁决的争端事项	219
函件 150	如果裁决人的裁决决定有利于我方,致雇主	220
函件 151	致雇主,请求共同任命仲裁人	221
函件 152	如果达不成共同任命仲裁人,致任命机构	223
函件 153	当工程实际竣工在即,致建筑师	224
函件 154a	当竣工证书无理由被扣压时,致建筑师	225
函件 154b	当竣工证书无理由被扣压时,致建筑师	227
函件 155a	致雇主,同意占有部分已竣工工程	228
函件 155b	致雇主,同意占有部分已竣工工程	229
函件 156	致雇主,要求签发占有部分已竣工工程的书面证明	230
函件 157	致雇主,不同意占有部分已竣工工程	231
函件 158	收到缺陷清单后,致建筑师	232
函件 159	当缺陷整改工作完成后,致建筑师	233
函件 160	致建筑师,在最后付款后返还图纸等文件材料	234

第九章　分包商与分包合同　　　　　　　　　　　　　　　　　235

函件 161	致雇主,请求同意权益转让	236
函件 162	当请求雇主同意权益转让时,致雇主	237
函件 163	致建筑师,请求同意分包	238
函件 164	致雇主,请求按第 19.3.1 条,在人员名单中增加人员	239
函件 165	致雇主,同意按第 19.3.1 条,在人员名单中增加人员	240
函件 166	致分包商:意向函	241
函件 167	致国内分包商,当招标邀请信中已提及时,要求提供担保	242
函件 168	致国内分包商,当招标邀请信中未提及时,要求提供担保	243
函件 169	当国内分包商不愿提供担保时,致建筑师	244
函件 170a	致建筑师,反对指定分包商	245
函件 170b	致建筑师,反对指定人员	246
函件 170c	致建筑师,反对指定分包商	247
函件 171a	当承包商不能与指定分包商达成协议或不能签订分包合同时,致建筑师	248
函件 171b	当承包商不能与指定人员签订分包合同时,致建筑师	250
函件 171c	当根据合同特殊条款不能与指定人员签订分包合同时,致建筑师	251
函件 172	当建筑师不能证实不一致时,致建筑师	252
函件 173	当指定的供应商不愿签订合适的采购合同时,致建筑师	253
函件 174	当所推荐的分包商撤回其报价时,致建筑师	254
函件 175	当承包商与指定分包商签订分包合同后,致建筑师	255
函件 176	当承包商与指定的人员签订分包合同后,致建筑师	256
函件 177	当承包商与指定的人员签订分包合同时,致建筑师	257
函件 178	关于保险事宜,致分包商	258
函件 179	当分包商不能办理保险时,致分包商	259
函件 180	致分包商,并附上图纸	260
函件 181a	当要求批准分包时,致指定分包商	261
函件 181b	当要求批准分包时,致分包商	262
函件 182a	当分包未获批准时,致指定分包商	263
函件 182b	当分包未获批准时,致分包商	264
函件 183a	致指定分包商,同意分包	265
函件 183b	致分包商,同意分包	266
函件 184	致分包商,要求服从指示	267
函件 185	当承包商不同意所谓的口头指令时,致分包商	268
函件 186	致指定分包商,附上建筑师关于授权条款的信函	269
函件 187	致建筑师,关于指定分包商工期延误的通知	270

函件 188a	致指定分包商,同意工期顺延	271
函件 188b	致指定分包商,同意工期顺延	272
函件 189	当不同意工期顺延时,致分包商	274
函件 190a	当工期顺延的索赔无效时,致指定分包商	275
函件 190b	当工期顺延的索赔无效时,致分包商	276
函件 191	当指定分包商不能按时竣工时,致建筑师	277
函件 192	当建筑师不同意按第 35.15 条规定签发证书时,致建筑师	279
函件 193	当未能按时完成工程时,致分包商	280
函件 194	关于分包工程实际竣工后审核工期顺延要求时,致建筑师	281
函件 195	致指定分包商,要求为费用索赔提供证据资料	282
函件 196	致指定分包商,要求提供损失和(或)费用的详细资料	283
函件 197	致分包商,要求提供费用索赔的进一步资料	284
函件 198	致分包商,要求偿付所蒙受的损失和(或)费用	285
函件 199	致分包商,关于期中付款通知	286
函件 200	致分包商,发出不支付工程款的通知	287
函件 201a	致指定分包商,并附上付款支票	288
函件 201b	致分包商,并附上付款支票	290
函件 202	致建筑师,附上已支付指定分包商工程款的证据	291
函件 203	当承包商不能提供已支付指定分包商工程款的证明时,致建筑师	292
函件 204	致裁决人,并附上书面说明	293
函件 205	当裁决人已聘用但并无争端时,致分包商	294
函件 206	无争端时,致裁决人	295
函件 207	当分包商已提前 7 天错误地发出要暂停履约的意向函时,致分包商	296
函件 208	当分包商提前 7 天正确地发出要暂停履约的意向函时,致分包商	299
函件 209	致分包商,请求调整或记录最终分包合同总价的文件资料	300
函件 210	致建筑师,附上指定分包商的文件资料	301
函件 211	当第 35.17 条的证书在所有分包工程缺陷整改完成前已签发时,致建筑师	302
函件 212	致分包商,附上最终付款	303
函件 213a	致建筑师,通知他指定分包商的违约	305
函件 213b	致建筑师,通知他指定分包商的违约	306
函件 214	当指定分包商不再违约时,致建筑师	307
函件 215	致分包商,在合同终止前通知其违约	308
函件 216	当指定分包商继续违约时,致建筑师	309
函件 217	当有可能终止对指定人员的雇用时,致建筑师	310
函件 218	致建筑师,寻求终止指定分包商的雇用的指令	311
函件 219a	致分包商,在发出违约通知后终止雇用关系	312

函件 219b	致分包商,终止雇用关系	313
函件 220a	当指定分包商继续违约,承包商终止分包合同时,致建筑师	314
函件 220b	当指定人员的雇用被终止时,致建筑师	315
函件 220c	当根据 NAM/SC 合同第 27.1 条或第 27.2 条已终止对指定人员的雇用时,致建筑师	316
函件 221	当要求承包商批准替代分包商的报价时,致建筑师	318
函件 222	当分包合同终止后,承包商被指令实施指定人员的工程时,致建筑师	319
函件 223	当指定人员的雇用已终止,承包商决定分包时,致建筑师	320
函件 224	当分包合同终止且费用已收回时,致雇主	321
函件 225	当指定分包商告知工程已实际竣工时,致建筑师	322
函件 226	致指定分包商,附上实际竣工时的记录	323
函件 227	当承包商不同意分包工程的实际竣工日期时,致分包商	324
函件 228	致分包商,附上缺陷清单	325
函件 229	致分包商,指出某些缺陷尚未整改完成	326
函件 230	当指定分包商设计失误而使承包商受到威胁时,致建筑师	328
函件 231	当指定人员的设计失误而使承包商受到威胁时,致建筑师	329
函件 232	关于职业责任保险,致分包合同的建筑师、工程师或其他工程顾问	330
函件 233	关于担保事项,致分包合同的建筑师、工程师或其他工程顾问	331
函件 234	未能及时提供资料时,致分包合同的建筑师、工程师或其他工程顾问	332
函件 235	当设计失误而使承包商受到威胁时,致分包合同的建筑师、工程师或其他工程顾问	334
函件 236	当工程项目顺利完成时,致分包合同的建筑师、工程师或其他工程顾问	335
名词解释		336

译者前言

我国加入世贸组织以后，随着我国建筑市场的逐步对外开放，我国建筑承包商正在努力学习并掌握工程项目实施的国际惯例，积极参与国际工程项目的竞争并逐步提高工程管理的整体水平，呈现出一派蒸蒸日上的大好景象。

工程管理实施过程中，业主、承包商与顾问工程师/建筑师（我国称为监理工程师）之间频繁的信函往来是一项十分重要但又十分繁琐的信息管理工作。我们深知，我国不少承包商不善于撰写这类信函，有的信函格式不规范，有的信函语言表达欠妥，更有少量的承包商不善于用书面形式反映问题，因而在合同索赔这一重要环节上吃了亏。

英国合同专家 David Chappell 先生编著的这本《建筑承包商标准信函格式》，包罗了承包商在履行合同的全过程中可能会出现的许多常见情况时应撰写的各类信函，共 270 多封。这些信函格式正确，语言规范，用词妥切，现以中英文对照形式出版，它是我国建筑施工承包商难得的一本好读物。在中国建筑工业出版社的大力支持下，本书的出版必将对我国从事国际工程项目管理的专业机构（监理公司，施工企业等）和从业人士十分有用，同时它也是我国高等院校相关专业的专业英语教学的一本很好的辅导读物。

正如原书作者在"绪论"中所告诫的，我国读者在使用本书时，切勿生搬硬套，在撰写每一封信函时，必须根据实际情况对信函范例进行修改。虽然这些信函都与当今在英国通用的各类标准合同文本有关，特别是信中引用的条款都是来自于那些标准合同条件。但只要稍做改动，这些信函标准格式均可适用于我国承包商熟知的国际上通用的 FIDIC 合同条件，英国土木工程师学会编制的 NEC 合同条件以及美国建筑师学会编制的 AIA 合同条件。

尽管我们几位译校者在译校过程中努力做到既忠实于原文又尽量符合中文表达习惯，历经数月，多次切磋，但由于时间仓促，再加上我们水平和经验有限，译文中不妥乃至错译之处在所难免，恳请各位同仁及广大读者不吝赐教，以便再版时更正。

参加翻译工作的有苏州科技学院方志达、余培明、周常明以及中建总公司上海公司夏瑞康。翻译分工如下：

周常明	第 1~3 章及第 8 章	函件	1~33, 139~160
夏瑞康	第 4 章	函件	34~99
余培明	第 5~7 章	函件	100~138
方志达	第 9 章及其他	函件	161~236

全书由方志达校对并定稿。

方志达
苏州科技学院　国际工程管理研究所
2003 年 12 月

第三版前言

十分高兴地获知本书仍在热销中。因为建筑施工合同和施工分包合同一直在不断修改,对许多条款进行逐一审查并非易事。十分希望,这一版本与前一版本一样简洁明了。

除了对所有信函进行逐一修正并根据法律和法规,取消了部分函件又增加了一些函件,内容仍有一些大的变化。本书现有270多封信函。"国内分包合同(DOM/1)"的信函已取消,而用新的"JCT国内分包合同标准格式(2002)"替代。但是,DOM/2合同目前仍保留使用。引入"房屋建筑与土木工程标准施工分包合同(GC/Works Sub-Contract)"与"房屋建筑与土木工程标准施工合同(GC/Works/1(1998))"配套使用。本书使用了JCT98、IFC98、MW98和WCD98标准合同的最新版本以及每种版本的修正条款1~4条,也使用了"房屋建筑与土木工程标准施工合同(GC/Works/1(1998))"的最新版本。

由于取消了部分函件又增加了一些函件,对一些说明作了修改。每一章的介绍说明仍保持最少篇幅,希望信函自身及其标题和注释能不释自明。

感谢我夫人玛格丽特在我完成这一版本的修改过程中给予的一贯支持。

大卫·查贝尔　David Chappell
于韦克菲尔德　Wakefield

绪论

本书是为承包商撰写的，然而作者也希望分包商能从中获益。本书是由一系列标准函件组成，这些函件与标准施工合同和标准分包施工合同配套使用。即使是一个小金额合同，只要采用一种标准合同格式，承包商都要为此撰写大量函件。有些函件是承包商必须发出的通知，有些则是经深思熟虑才会发出的函件。为特定情况撰写函件是件十分乏味的工作。对于同一种情况，不时地重复撰写同一种内容的信，既费时又费力，而且浪费钱财。

本书试图包罗承包商在履行合同中遇到的所有常见的情况。当然，要做到这一点难度很大，有些情况难免挂一漏万。作者将非常欢迎读者通过出版商就本书中的标准函件提出宝贵意见，以便今后再版时修改。

在使用标准或范例函件时，人们常见的批评是，读者无视不同情况套用这些标准的函件。尽管会出现这种风险，但这只能归咎于读者的粗心大意。正是这种粗心大意的出现，才导致他们写出不恰当的函件。人们提出这样的批评，实际上是因为读者不明白写这类函件的目的是什么。读者有必要了解合同的内容，然后对函件的内容做出判断。我希望书中的函件范例能够成为读者撰写函件时的对照表，成为函件内容的撰写指南和在特定情况下遣词造句的示范。

书中没有对诸如[填入日期]一类的字样做过多的解释。如果在函件中加上假设的日期，使用假设的公司名称，列举假设的情景说明各种情况，也是可以的。如果这样做了，可能书写起来更有些趣味。我不想批评这种做法，尽管这种做法在其他场合下会带来较好的效果，但是，在我看来，这种方法在使用中似乎会造成更多的出错可能。

为了把大量的函件缩减到可以分类管理的篇幅，所有的函件被分成若干部分。在这本书的姊妹篇《建筑实践标准信函格式》中，英国皇家建筑师协会编制的工程分类规划常被作为函件序列的编写框架。这种做法在本书情况下是不合适的，因为承包商的工作通常也不是按照该分类进行的。

相反，书中各章节的标题是按大量的重要活动的先后顺序排列，即：投标，合同文件，保险，现场作业，付款，延期，损失与费用，合同终止，仲裁、裁决与工程竣工，以及分包商。在每一部分中，函件是按照最有可能被用到的第一种情况的基础上编排的，因此，查找任何一种函件都是很容易的。为了便于查找，本书附有关键词索引，作为内容总目录的补充。

除非每封函件开头另有说明，所有函件都适用于标准施工合同格式（JCT98），承包商负责设计的标准施工合同格式（WCD98），中等规模标准施工合同格式（IFC98），小型建筑工程协议（MW98），房屋建筑与土木工程标准施工合同的一般条件（GC／Works／

1(1998))。当不同的标准合同要求须撰写不同的函件时,均有注释加以说明。尽管本书为了总体上简洁明了,每封函件只涉及单独的一个主题,可以理解的是,在实际应用中,多个主题常常涵盖在一封信中。

关于分包商的部分因篇幅最长,编排时放在本书的最后。该章节介绍了分包合同的各种格式,反映的活动与总合同的活动相仿。该章节中包含了致分包商的函件以及就分包合同问题致建筑师的函件。分包合同包括:指定分包合同标准格式(NSC/C);适用于根据中等规模标准施工合同指定的分包商的分包合同条件标准格式(NAM/SC);2002年国内分包合同(DSC/C)JCT标准格式;国内分包合同标准格式(DOM/2)以及2000年房屋建筑与土木工程标准施工合同的分包合同标准格式(GC/Works/SC)。本书中涉及的主合同与分包合同之间的关系如下:

JCT98	NSC/C
	DSC/C
WCD/98	DOM/2
IFC98	NAM/SC
MW98	– – –
GC/Works/1	GC/Works/SC

编写本书之际,还没有与MW98合同配套的分包合同的标准格式。

在使用本书时,应记住以下几点:

● 当提及JCT98合同时,主要指带工程量清单的私人项目合同。尽管大多数函件可适用于其他五种合同格式,但在使用前应谨慎核对,因为在某些情况下还是有差别的。

● 每一份函件都应该有一个标题,以表示信函的内容。但是,为简明扼要起见,书中的标题一概省略。

● 为方便文笔,书中一律使用阳性的"他"。但是"他"有时可用来表示"她";同样,"他的"也可用来表示"她的"。

● 为适应不同的合同类型,在同一封函件中内容都有些变化。但为了更方便,少混淆或者突出重要函件,函件都会单独另立,并在函件编号上增加a、b、c、等符号以示区别。

● 书中通篇采用"建筑师"、"委托人"、"雇主"等称呼语,但应该指出,当采用GC/Works/1(1998)合同或者GC/Works/SC合同时,雇主称作"行政当局",建筑师称作"项目经理"(PM);当采用WCG98合同时,传统上没有建筑师这一角色,写给"建筑师"的函件通常应发给指定的"雇主代表",否则就给"雇主"。

● 人们一直认为,补充条款应与WCD98合同配套使用。

● 从其他途径能够得到的标准文件,诸如投标函标准格式和各种证书格式均未收入本书。

● 承包商可能要撰写的有关增值税和财政法案以及有关物价波动等问题的函件没有收入本书。因为这些情况往往会发生突变和(或)在多数情况下不适合用标准函来处理。

应引起警觉的是:对于一个业务繁忙的承包商来说,使用标准函件是十分有用的。但是,如不加思考就照搬,则是非常危险的。在使用标准函件时,一定要认真考虑该函件是否真正适合于所表达的特定情况。有疑问时,就应寻求帮助。

第一章　招投标

　　在施工招投标程序中常用的各种标准格式已出版。以下的函件是基于采用1996年单一阶段选择性招标程序规范而编写的。如果采用1996年两阶段选择性招标程序规范或1996年设计和建造合同选择性招标程序规范，这些函件须做相应改动。

　　通常，投标阶段是承包商与建筑师之间的首次接触。这个阶段既漫长，又令人心碎，尤其在错过了提交投标文件截止日期的时候，而这种情况也是屡见不鲜的。要求承包商按照WCD98合同参与设计和建造总承包工程项目的投标，其投标期往往会比按照传统采购模式所需的时间长得多。这是因为承包商在着手准备报价之前必须要有额外的时间来提出设计方案。为了避免时间和金钱的严重浪费，通常的做法是分两个阶段进行招标。

　　投标是承包商获得工程的途径。承包商应抱着耐心的态度，克服招标初期的困难，力争列入被邀请投标的名单之中。许多地方行政管理部门都拥有承包商的名单。关于哪些公司可被列入邀请投标的名单中，私有雇主往往向建筑师和工料测量师咨询。当然，最后是由雇主决定。第一封信就是为这种情况而设计的，即你得知有一个项目正处在设计阶段中。虽然这种函件是致建筑师的，但在有些情况下，例如你觉得建筑师可能不愿意帮忙的话，写给雇主也行。如果寄出的函件没有回音，继续努力去跟踪总是值得的。

函件 1
关于要求列入邀请投标人名单,致建筑师

Letter 1
To architect, requesting inclusion on list of tenderers

尊敬的先生:

  我们有幸从[注明消息来源]注意到上述项目近期要进行招标。对于这类项目我们十分富有经验,如蒙邀请参加投标,我们将不胜感激。若你方对我方的施工能力不甚了解,以下信息或许对你方很有帮助。

[列出以下信息:
 1. 公司全体董事的姓名和地址
 2. 注册办公地点
 3. 网址
 4. 公司的股份资产
 5. 过去三年的年产值
 6. 办公室职员的人数及职务
 7. 每个工种永久雇用的现场施工人员人数
 8. 经过培训的管理人员人数
 9. 目前在建合同的数量和造价
10. 三个近期竣工且与招标项目特点相似的工程项目的地址,竣工日期和工程造价
11. 能够提供证明的上述第10条中注明的雇主、建筑师和工料测量师的姓名和地址]

企盼着你方的回音。

你忠诚的

Dear Sir

We were interested to note from [*state source*] that tenders are to be invited for the above project in the near future.

This is a type of work in which we are very experienced and we should welcome an invitation to tender. The following information may be of assistance to you if you are not already aware of our capabilities:

[*List the following information*:

1. Names and addresses of all directors.
2. Address of registered office.
3. Website.
4. Share capital of firm.
5. Annual turnover during the last three years.
6. Number and positions of all office-based staff.
7. Number of site operatives permanently employed in each trade.
8. Number of trained supervisory staff.
9. Number and value of current contracts on site.
10. Address, date of completion and value of three recently completed projects of similar character to that for which tenders are to be invited.
11. Names and addresses of clients, architects or quantity surveyors connected with the projects noted in 10 above and to whom reference may be made.]

We look forward to hearing from you in due course.

Yours faithfully

函件 2
关于要求列入邀请投标人名单,但未有回音,致建筑师

Letter 2
To architect, if no response to request for inclusion on list of tenderers

---

尊敬的先生:

  谨提及我方于[填入日期]发出的有关要求列入上述项目邀请投标人名单的函。
  由于我方尚未获悉你方的答复,故借此机会再次表明我方对该项目的兴趣,并向你方确认,我方对此类工程的施工很有经验。
  我方衷心希望能与你方会晤,并就上次信件中提及的细节问题进行详细讨论。我们将用设备来演示我方近期完成的工程项目,相信你方雇主会感兴趣。
  我方总经理 [或填入相关头衔],[填入姓名]先生,将于 [填入日期] 与你电话联系。

你忠诚的

Dear Sir

We refer to our letter of the [*insert date*], requesting inclusion on the list of tenderers for the above project.

Since we have not heard from you, we take this opportunity to re-affirm our interest in the project and assure you of our experience in work of this nature.

We should be delighted to meet you to expand upon the details given in our earlier letter. We have the facilities to make a presentation showing recent projects we have carried out which may be of interest to your client.

Our managing director [*or insert appropriate designation*], Mr [*insert name*], will telephone you on [*insert day*].

Yours faithfully

函件 3
同意投标,致建筑师

Letter 3
To architect, agreeing to tender

尊敬的先生：

  你方于[填入日期]的来函收悉,谢谢。从中,我方得知你方打算就上述工程项目进行招标。

  若我方被列入邀请投标人名单,我方将深感荣幸。相信你方将会很快寄来详细的资料。

你忠诚的

Dear Sir

Thank you for your letter of the [*insert date*] from which we note that you intend to invite tenders for the above project.

We should be pleased to be included on the tender list. No doubt you will be sending further details in due course.

Yours faithfully

函件 4

如果承包商不愿投标,致建筑师

Letter 4

To architect, if contractor unwilling to tender

尊敬的先生:

　　你方于[填入日期]的来函收悉。我方注意到你方打算为上述工程项目招标。

　　我方遗憾地通知你方,由于我方日前工程任务繁重,无法参加你方的招标,谨请原谅。然而,我们真切地希望将来有机会参与你方其他工程项目的投标。

你忠诚的

Dear Sir

Thank you for your letter of the [insert date] from which we note that you intend to invite tenders for the above project.

With regret, we must ask to be excused from tendering on this occasion due to our very heavy workload. We do hope, however, that you will give us the opportunity to tender for other projects on other occasions in the future.

Yours faithfully

函件 5
如果在被列入邀请投标人名单之前要求承包商提供材料,致建筑师

Letter 5
To architect, if contractor asked to provide information prior to inclusion on tender list

尊敬的先生:

　　收到你方 [填入日期] 来函,介绍了上述工程项目的简要情况以及要求了解本公司详细情况。
　　根据你方提供的资料,该项目与我们现有的施工技术和经验密切相关。若能被列入邀请招标人名单,我方将倍感荣幸。你方要求的有关资料如下:

[根据建筑师信函中提出问题的序号,逐项给予答复]

你忠诚的

Dear Sir

Thank you for your letter of the [*insert date*] setting out brief details of the above project and requesting particulars of this company.

On the information provided, it appears that the project would be directly related to our skills and experience and we should be delighted to be included on the tender list. The information you require is as follows:

[*List answers using same numeration as in architect's letter*]

Yours faithfully

函件 6
如果通知承包商投标日期推迟,承包商仍愿参与投标时,致建筑师

Letter 6
To architect, if the contractor is informed that the tender date is delayed and is still willing to submit tender

尊敬的先生:

　　你方于[填入日期]来函通知,投标文件发放日期已更改至[填入日期]。我方确认,仍愿意参与该项目投标。

你忠诚的

Dear Sir

Thank you for your letter of the [*insert date*] informing us that the date for despatch of tender documents has been revised to [*insert date*]. We confirm that we are still willing to submit a tender for this project.

Yours faithfully

函件 7
如果承包商得知投标日期被推迟,且不愿意再参与投标时,致建筑师

Letter 7
To architect, if the contractor is informed that the tender date is delayed and is unwilling to tender

尊敬的先生:

　　你方于[填入日期]通知我方,投标文件的发放日期已更改至[填入日期]的来函收悉。由于我方无法重新调整繁重的工程任务,因而,无法按照新的时间表参加你方的招标,对此深表遗憾。
　　然而,我们真切地希望将来有机会参与你方其他工程项目的投标。

你忠诚的

Dear Sir

Thank you for your letter of the [*insert date*] informing us that the date of despatch of tender documents has been revised to [*insert date*]. We regret that it will be impossible for us to rearrange our very heavy workload so as to be able to submit a tender in accordance with the new timetable.

We do hope, however, that you will give us the opportunity to tender for other projects on other occasions in the future.

Yours faithfully

函件 8

确认收到投标文件,致建筑师

Letter 8

To architect, acknowledging receipt of tender documents

尊敬的先生:

　　十分感谢你方正式邀请我方参加上述工程项目的投标,并确认收到下列附件:[列出附函文件]

　　我方确认,将于[填入日期]之前提交投标书。
　　[视情形,添入以下内容:]

　　我方希望认真研究图纸并考察施工现场。我方[填入姓名]先生将于近日与你电话联系,以确定必要的会面。

你忠诚的

Dear Sir

Thank you for your formal invitation to tender for the above project with which you enclosed [*list documents enclosed*].

We confirm that we will submit our tender by the [*insert tender date*].

[*If appropriate, add*:]

We wish to inspect the detailed drawings and visit site. Our Mr [*insert name*] will telephone you to make the necessary appointment within the next few days.

Yours faithfully

函件 9
关于投标期间的质疑,致建筑师

Letter 9
To architect, regarding questions during the tender period

尊敬的先生:

我方仔细研究了你方附在[填入日期]函件内的招标文件。在你方办公室查阅了工程详图,并考察了施工现场,我方仍有以下问题待澄清:

[列出需要澄清的事项]

标有红色"X"符号的为紧急事项。如果要按时提交投标书的话,我方需要在[填入日期]之前得到答复。

你忠诚的

Dear Sir

We have carefully examined the tender documents enclosed with your letter of the [*insert date*]. We have examined the detailed drawings at your office and we have visited site. There are certain items which require clarification as follows:

[*List items requiring clarification*]

Items marked with a red X are urgent and, if we are to meet the date for submission of tenders, we need clarification of these points by [*insert date*].

Yours faithfully

函件 10
关于要求延长投标期限,致建筑师

Letter 10
To architect, requesting extension of tender period

尊敬的先生:

  我方正以最快的速度为上述项目准备报价。然而,不少分包项目的报价要等到投标日期之后,我方才能收到。很明显,除非延长投标期限,否则我们无法按期提交标书。因此,我方要求将投标期限延长一周。在假设你方会同意我方要求的前提下,我方会继续完成投标工作。但如果你方觉得无法做到延期,请即告知,以便我方减少无效工作。

你忠诚的

Dear Sir

We are preparing our tender for the above project with the greatest possible speed. Prices for a number of the sub-contract items, however, will not be in our hands until after the date for submission of tenders. Clearly, we will be unable to submit a tender unless the tendering period is extended. We therefore request an extension of the period by one week. We are proceeding on the assumption that you will be able to grant our request, but if you feel unable so to do, please let us know immediately so that we can reduce what will become abortive work.

Yours faithfully

函件 11
撤回投标资格,致建筑师

Letter 11
To architect, withdrawing qualification to tender

尊敬的先生:

  为答复你方[填入日期]的来函,我方确认,我方撤回于[填入日期]的投标资格并无意更改标书中的标价[填入金额]。

  上述提及的投标资格是指:

[用标书中使用的准确措词陈述该资格]

你忠诚的

Dear Sir

In response to your letter of the [*insert date*], we confirm that we withdraw the qualification to our tender dated [*insert date*] without amendment to the tender sum of [*insert amount*].

The qualification to which we refer above is:

[*Set out the qualification using the precise wording used in the tender*]

Yours faithfully

函件 12
致建筑师,根据方案 1 或 2 确认报价

Letter 12
To architect, if confirming offer in accordance with Alternative 1 or 2

尊敬的先生:

  收到你方于 [填入日期] 的来函并附有一份有关我方上述项目工程量表报价中的错误清单,谢谢。

  我方仔细审查了该清单。根据1996年单阶段选择性招标程序规范中第六部分的方案 1 或 2,请将本函作为通知,我方确认我方于[填入日期]提交的投标价总额为[填入金额]。

  我方注意到你方关于背书的说明,我方同意这些条件。

你忠诚的

Dear Sir

Thank you for your letter of the [*insert date*] with which you enclosed a list of errors detected in our pricing of the bills for the above project.

We have carefully examined the list and, in accordance with Alternative 1/2 [*delete as appropriate*] of section 6 of the Code of Procedure for Single Stage Selective Tendering 1996, please take this as notice that we confirm our offer of [*insert amount*] as stated in our tender dated [*insert date*].

We note what you state regarding endorsement and we agree to its terms.

Yours faithfully

函件 13
根据方案1,要求撤回报价时,致建筑师

Letter 13
To architect, if withdrawing offer in accordance with Alternative 1

尊敬的先生:

收到你方于[填入日期]的来函并附有一份有关我方对上述项目工程量表报价中的错误清单,谢谢。

我方仔细审查了该清单。考虑到这些错误的性质,根据1996年单阶段选择性投标程序规范中第六部分的方案1,我方决定撤回报价,并对此深表遗憾。

你忠诚的

Dear Sir

Thank you for your letter of the [*insert date*] with which you enclosed a list of errors detected in our pricing of the bills for the above project.

We have carefully examined the list and, in view of the nature of the errors, we regret that we must withdraw our offer in accordance with Alternative 1 of section 6 of the Code of Procedure for Single Stage Selective Tendering 1996.

Yours faithfully

函件 14
根据方案 2,要求修改报价时,致建筑师

Letter 14
To architect, if amending offer in accordance with Alternative 2

尊敬的先生:

  收到你方于[填入日期]的来函并附有一份有关我方对上述项目工程量表报价中的错误清单,谢谢。

  我方仔细审查了该清单。考虑到这些错误的性质,根据 1996 年单阶段选择性投标程序规范中第六部分的方案 2,我方已修改了报价,以改正错误。我们修改后的报价为[填入金额],并附上相关的计算书。

你忠诚的

Dear Sir

Thank you for your letter of the [*insert date*] with which you enclosed a list of errors detected in our pricing of the priced bills for the above project.

We have carefully examined the list and, in view of the nature of the errors, we have amended our offer, in accordance with Alternative 2 of section 6 of the Code of Procedure for Single Stage Selective Tendering 1996, to correct the errors. Our amended tender price is [*insert amount*] and we enclose details of the relevant calculations.

Yours faithfully

函件 15a
如果投标书被接受，致建筑师
本函不适用于 WCD98 合同

Letter 15a
To architect, if tender accepted
This letter is not suitable for use with WCD 98

尊敬的先生：

　　谢谢你方[填入日期]的来函并接受了我方于[填入日期]提交的投标书。根据编号为[填入编号]的图纸和工程量清单[或技术规范]，投标价为[填入金额]。

　　我方深知，现在雇主与我们之间建立了合同关系，并期待能尽快收到作为契约进行签署/履行[视情形取舍]的合同文件。

你忠诚的

Dear Sir

Thank you for your letter of the [*insert date*] accepting our tender of the [*insert date*] in the sum of [*insert amount*] for the above work in accordance with the drawings numbered [*insert numbers*] and the bills of quantities [*or specification*].

We understand that a contract now exists between the employer and ourselves and we look forward to receiving the contract documents for signing/execution as a deed [*delete as appropriate*] in due course.

Yours faithfully

函件 15b

如果投标书被接受,致建筑师
本函仅适用于WCD98合同

Letter 15b

To architect, if tender accepted
This letter is only suitable for use with WCD 98

尊敬的先生:

　　谢谢你方[填入日期]的来函,接受了我方于[填入日期]提交的完成上述工程项目的设计和施工的标价为[填入金额]的投标书。该报价是根据雇主的要求、承包商的建议以及合同总价分析编制的。

　　我方深知,现在雇主与我们之间建立了合同关系,并期待能尽快收到作为契约进行签署/履行[视情形取舍]的合同文件。

你忠诚的

Dear Sir

Thank you for your letter of the [*insert date*] accepting our tender of the [*insert date*] in the sum of [*insert amount*] for the completion of the design and the construction of the above project in accordance with the Employer's Requirements, the Contractor's Proposals and the Contract Sum Analysis.

We understand that a contract now exists between the employer and ourselves and we look forward to receiving the contract documents for signing/execution as a deed [*delete as appropriate*] in due course.

Yours faithfully

函件 16
如果其他承包商的投标书被接受时,致建筑师

Letter 16
To architect, if another tender accepted

尊敬的先生:

你方于[填入日期]的来函收悉。从中得知,我方此次投标未中标。

我方期待能看到一份完整的各家投标价清单,并希望回答你方进一步的问题。

你忠诚的

Dear Sir

Thank you for your letter of the [*insert date*] from which we note that our tender was not successful in this instance.

We await details of the full list of tender prices with interest and assure you of our willingness to receive your future enquiries.

Yours faithfully

第二章　合同文件

　　如果你投标成功,建筑师或雇主将会写信通知你。这类信件可以是(也可以不是)一份具有约束力的成为合同文件之一的中标函。一封措词明确的中标函当然是你所期盼的,但是,这是不常见的。如果信件中设置了附带条件或就某些事项需进一步达成一致的条款,通常这种信件就不具备合同效力。

　　本章中的函件涉及合同文件中可能会出现的一些基本情况,这些合同文件是准备作为契约由建筑师签署或完成,或涉及当地行政管理部门,则由法定部门签署或完成。现在已无必要非在合同文件上盖章才能使之成为有效契约了,但是如果你还想那么做,仍可照旧行事。鉴于合同文件是意向还是成为契约,其合同文件适用的限期有所不同(可能是6年或12年),如果你对合同文件程序或后果存有疑虑,那么明智的做法是去咨询。但无论如何,关键的是要非常仔细地研读合同文件。

　　校对正式的合同文本时,应对照招标文件中提供的信息以及注明的差异之处进行逐一核对。合同图纸必须与你投标所依据的图纸一致。有时,在传统的采购合同模式情况下,在发出招标邀请到合同文件完成期间,建筑师会修改这类图纸。如果发现有任何不一致的情况,无论建筑师怎样向你承诺这些差异不会影响合同,应该拒绝签署这类合同文件,直至图纸修改完成为止。当使用 WCD98 合同时,尤其要谨慎。"雇主要求"文件和"承包商建议书"文件必须一致,因为合同本身没有任何可以纠正这两份文件之间差异的机制。还会有这样的风险,即在正式合同文件形成之前,作为你投标所依据的最初的图纸可能会被修改,而需要承担这类风险的人最有可能的是你自己,而不是雇主的代表。仅仅列出"雇主要求"和"承包商建议书"文件之间的差异是不够的,这些差异必须进行修正,达到一致才行。否则,"雇主要求"文件通常将会得到优先考虑。

　　有些函件涉及这类情况,即合同文件尚未签署之前工程就开工的情况。这种情况是不允许发生的,但在实际工作中却常常发生。如果一份有约束力的合同已谈妥并且该合同能准确反映正式合同文件的各项条款,这样做就不会出现什么麻烦。这一点通常可以通过添加条款的办法做到,即出现什么情况,就添加什么条款处理。只要把这些条款都添加进合同,那么,建筑师似乎就没有理由不签发诸如指令和证书了。

　　假如签发付款证书时间已到,不是一定要撰写所建议的那类函件,而是应该坚持在进驻工地现场之前使合同文件有效。在你的投标书被接受后,几乎没有什么理由不马上准备开工。但是,要记住,建筑师未能准备好合同文件不能成为你拒绝接收场地的理由,尤其是当一份合适的具有约束力的并且包含正式合同文件的所有条款的合同已形成时。

　　各种担保保证文件可以写成(已经是)一本书了,这里所附的函件仅适用于对要求签署担保协议函的几种答复。只有投标所依据的文件中有担保要求的条款时,你才有必要履行担保义务。

函件 17
寄还合同文件时,致建筑师
专递/挂号邮件

Letter 17
To architect, returning contract documents
Special / recorded delivery

尊敬的先生：

你方[填入日期]附有合同文件的来函收悉,谢谢。按照你方要求,我方将其作为契约如期签署/履行[视情形取舍],现将合同文件随信寄还。

我方期待着能在近日内收到正式签字的合同文件。

你忠诚的

Dear Sir

Thank you for your letter of the [*insert date*] with which you enclosed the contract documents. We have pleasure in returning them herewith, duly signed/executed as a deed [*delete as appropriate*] as requested.

We look forward to receiving a certified copy of the contract documents within the next few days.

Yours faithfully

函件 18
如果合同文件中出现错误,致建筑师

Letter 18
To architect, if mistakes in contract documents

尊敬的先生:

你方于 [填入日期] 随函所附的要求我方将其作为契约如期签署/履行 [视情形取舍] 的合同文件收悉。

合同文件中出现一个错误[说明错误的性质和错误所在的文件页码]。这与我方投标所依据的招标文件不相符,而我方的标书及你方于[填入日期] 发出的正式中标函已构成了雇主与我方之间具有约束力的合同。

为此,现随函退还这些文件,并期望尽快收到修正后的合同文本。

你忠诚的

Dear Sir

We are in receipt of your letter of the [*insert date*] with which you enclosed the contract documents for us to sign/execute as a deed [*delete as appropriate*].

There is an error on [*describe nature of error and page number of document*]. This is not consistent with the tender documents on which our tender is based which, together with your acceptance of the [*insert date*] forms a binding contract between the employer and ourselves.

We therefore return the documents herewith and we look forward to receiving the corrected documents as soon as possible.

Yours faithfully

合同文件

函件 19
如果合同文件中出现错误,而先前的投标书未被接受,致建筑师

Letter 19
To architect, if mistakes in contract documents and no previous acceptance of tender

尊敬的先生:

你方于[填入日期]随函所附的要求我方将其作为契约如期签署/履行[视情形删除]的合同文件收悉。

合同文件中出现一个错误[说明错误的性质和错误所在的文件页码]。这与我方投标所依据的投标文件不相符,因此,我们不打算依据目前的合同文件签订这份合同。

为此,现随函退还这些文件,并期望尽快收到修正后的合同文本。

你忠诚的

Dear Sir

We are in receipt of your letter of the [*insert date*] with which you enclosed the contract documents for us to sign/execute as a deed [*delete as appropriate*].

There is an error on [*describe nature of error and page number of document*]. This is not consistent with the tender documents on which our tender is based and we are not prepared to enter into a contract on the basis of the contract documents in their present form.

We therefore return the documents herewith and we look forward to receiving the corrected documents as soon as possible.

Yours faithfully

函件 20
如果要求承包商在合同文件签署之前开工,致建筑师

Letter 20
To architect, if contractor asked to commence before contract documents signed

尊敬的先生:

　　你方于[填入日期]的来函收悉,谢谢。我方注意到,雇主要求我们在合同文件签署完成之前就开始施工。

　　我方对这一情况的理解是:基于我方于[填入日期]递交的投标书以及雇主于[填入日期]正式发出的接受我方投标书所包含内容的中标函,我方已经与雇主处于一种具有约束力的合同关系之中。

　　雇主如能以书面形式发函,确认同意本函中我方的见解,我方将很高兴地按要求开工。

　　　　　　　　　　　　　　　　　　　　　　　　你忠诚的

Dear Sir

Thank you for your letter of the [*insert date*] from which we note that the employer requests us to commence work on site pending completion of the contract documents.

It is our understanding of the situation that we are already in a binding contract with the employer on the basis of our tender of the [*insert date*] and his acceptance of the [*insert date*] on terms incorporated by such tender and letter of acceptance.

If the employer will send us written confirmation that he agrees with our understanding of the situation as expressed in this letter, we will be happy to commence as requested.

Yours faithfully

函件 21 如果合同尚未签订且付款证书签证期已到,致建筑师 本函不适用 WCD98 合同 专递/挂号邮件 **Letter 21** To architect, if contract not signed and certification due *This letter is not suitable for use with WCD 98* *Special / recorded delivery*

尊敬的先生:

　　我方于[填入日期]以总价为[填入金额]投标承包上述工程。雇主于[填入日期]致函我方明确表示接受我方的投标书。在我方和雇主之间的有约束力的合同条件[填入合同形式的全称]应包含签证工程款付款证书等的合同条款。

　　尽管我方难以理解,你方为何至今尚未备好作为契约用于签署/履行[视情形删除]的正式合同文件,但是,完成签署这样的正式合同文件将直接反映出双方已达成的各自的权利和义务,而这一点是无庸置疑的。

　　因此,根据合同条件第 30 条[当使用 IFC98 合同或 MW98 合同时,为"第 4.2 条";当使用 GC/Works/1(1998)合同时,为"第 50 条"],我方要求你方签发付款证书。

　　如果到[填入日期]我方仍未收到付款证书,我方将立即采取法律行动起诉雇主的违约行为。

你忠诚的

Dear Sir

We tendered for the above work on the [*insert date*] in the sum of [*insert amount*] and the employer accepted our tender unequivocally by his letter of the [*insert date*]. A binding contract exists between the employer and ourselves in terms incorporating, among other things, the provisions of [*insert the full title of the form of contract*] which provides for certification of monies due to us.

Although we have difficulty in understanding why you have not yet prepared the formal contract documents for signature/completion as a deed [*delete as appropriate*], the completion of such formal contract documents will simply reflect the respective rights and obligations of the parties as already agreed and about which there is no doubt.

We, therefore, require you to issue your certificate in accordance with clause 30 [*substitute* "4.2" *when using IFC 98 or MW 98, or* "50" *when using GC/Works/1 (1998)*] of the conditions of contract

If such certificate is not in our hands by [*insert date*] we will take immediate legal action against the employer for the breach.

Yours faithfully

函件 22
如果要求承包商签署担保书,致建筑师

Letter 22
To architect, if contractor asked to sign a warranty

尊敬的先生:

收到你方于[填入日期]的来函及所附的一份待签字的担保书,谢谢。

[填入以下内容:]
我方注意到,这份担保书与附于合同文件中的担保书相同,为此,我方按要求把已签署完成的文本寄还给你方。

[或者填入以下内容:]
尽管合同文件要求我方完成担保书,但没有说明使用哪种担保书形式。遗憾的是,你方建议的担保书形式是不能接受的。

[或者填入以下内容:]
我方注意到,合同文件并无要求我方签署担保书的条款。然而,如果你方能在我方附上的担保书上签字并返给我方的话,那么,我方将准备签署所附上的建议的担保书。

你忠诚的

Dear Sir

Thank you for your letter of the [*insert date*] with which you enclosed a form of warranty for signature.

[*Add either*:]

We note that the form is identical to that attached to the contract documents and we have pleasure in returning it duly completed as requested.

[*Or*:]

Although the contract documents call on us to complete a form of warranty, no particular form was specified and we regret that your suggested form is not acceptable.

[*Or*:]

We note that there was no requirement in the contract documents for us to complete a warranty. However, we enclose a suggested warranty which we would be prepared to complete if you will complete a warranty to us as the form attached.

Yours faithfully

合同文件　　29

函件 23
如果要求承包商提供履约保函,致建筑师

Letter 23
To architect, if contractor asked to supply a performance bond

尊敬的先生:

　　你方于[填入日期]要求我方提供履约保函的来函收悉,谢谢。
　　[填入以下内容:]

　　我方知道,合同文件中附有履约保函格式,我方正在办理中。
　　[或者填入以下内容:]

　　尽管合同文件要求我方提供履约保函,但并没有规定保函格式。我们不能接受你方来函中所附的履约保函格式。我方正以一种可接受的格式办理保函。
　　[或者填入以下内容:]

　　合同文件并未要求我方提供履约保函,因此,我方不准备办理。

你忠诚的

Dear Sir

Thank you for your letter of the [*insert date*] requesting us to provide a performance bond.

[*Add either*:]

We note that a form of bond was included in the contract documents and we are arranging for a bond to be executed accordingly.

[*Or:*]

Although the contract documents call on us to provide a performance bond, no particular form was specified. The form you include with your letter is not acceptable and we are arranging to have the bond executed in an acceptable form.

[*Or:*]

The contract documents do not require us to provide a performance bond and we decline to do so.

Yours faithfully

第三章　保险

本章所提供的函件涉及到标准合同文本中所有的保险条款。除 GC/Works/1 合同外，其他合同的保险条款都极为相似。正如大家期望的，MW98 合同中的保险条款比其他 JCT 合同中保险条款要简短。应注意合同文本之间这些细微差异。如果已投保了普通性质的保险，只要这类保险所保障的风险和保障金额足够，这类保险就可以接受。即使承包商的工程一切险按年投保，那也必须确保至少把每一新项目向保险人通报。保险是一门专业，需要保险经纪人认真对待。

如果雇主投保，例如，对现有的建筑物和扩建工程投保，务必在进场之前了解保险情况。一般不要求当地政府机构投保，可能是因为利用其自己的资源应对风险会更经济。

自 1980 年最后一次对 JCT 合同做重大修改后，对保险条款的修订从没中断过，明显的是因为考虑了法庭对案例的判决结果。

函件 24a
有关承包商的保险，致建筑师
本函不适用于 MW98 或 GC/Works/1(1998)合同
专递/挂号邮件

Letter 24a
To architect, regarding contractor's insurance
This letter is not suitable for use with MW 98 or GC/Works/1(1998)
Special/recorded delivery

尊敬的先生：

你方[填入日期]的来函收悉。应你方要求，根据合同条件第21.1.2条[使用IFC98合同时，为"第6.2.2条"]规定，现附上下列保险单及保险费收据，请雇主过目。请寄回所有用专递或挂号邮递给你方的单据，并附函说明雇主对上述单据无疑义。

[根据序号和日期列出保险单和保险费收据清单]

[视情形，可添入以下内容：]
根据合同条件第22A.2条[使用IFC98合同时，为"第6.3A.2条"]规定，现附上我方建议的保险人[填入姓名和地址]情况，请雇主核准。

[或：]
根据合同条件第22A.3.1条[使用IFC98合同时，为"第6.3A.3.1条"]规定，现附上证明文件，表明依据本合同我方已履行投保工程一切险的义务。请注意，附件中所填的内容：每年的续保日期为[填入日期]。

你忠诚的

Dear Sir

Thank you for your letter of the [*insert date*]. In response to your requirement, we have pleasure in enclosing the following insurance policies and premium receipts in accordance with clause 21. 1. 2 [*substitute* "*6. 2. 2*" *when using IFC 98*] of the contract for inspection by the employer. Please return all the enclosed documents by special or recorded delivery with a note stating that they are to the employer's satisfaction.
[*List policies and receipts together with numbers and dates*]

[*If appropriate, add either*:]

In accordance with clause 22A. 2 [*substitute* "*6. 3A. 2*" *when using IFC 98*], we hereby submit for the employer's approval [*insert name and address*] with whom we propose to take out insurance.

[*Or*:]

In accordance with clause 22A. 3. 1 [*substitute* "*6. 3A. 3. 1*" *when using IFC 98*] we enclose documentary evidence that we maintain appropriate All Risks cover independently of our obligations under this contract. Please note, for insertion in the appendix, that the annual renewal date is [*insert date*].

Yours faithfully

函件 24b

有关承包商保险，致建筑师

本函仅适用于 MW98 合同

专递/挂号邮件

Letter 24b

To architect, regarding contractor's insurance

This letter is only suitable for use with MW 98

Special / recorded delivery

尊敬的先生：

　　你方[填入日期]的来函收悉。应你方要求，根据合同条件第 6.4 条规定，现附上下列保险单及保险费收据，请雇主过目。

　　[根据序号和日期列出保险单和收据清单]

　　[视情形，可添入以下内容：]

　　同时，附上下列保险单及保险费收据以确认我方依据本合同条件第 6.3A 条规定已办理了有关保险。

　　[根据序号和日期列出保险单和保险费收据清单]

　　[或：]

　　现附上证明文件，以证明依据本合同我方已履行了投保工程一切险的义务。

　　[然后，添入以下内容：]

　　请寄回所有用专递或挂号邮递给你方的单据并附函说明雇主对上述单据无疑义。

你忠诚的

Dear Sir

Thank you for your letter of the [*insert date*]. In response to your requirement, we have pleasure in enclosing the following insurance policies and premium receipts in accordance with clause 6.4 of the contract for inspection by the employer.
[*List policies and receipts together with numbers and dates*]

[*If appropriate, add either*:]

We also enclose the following insurance policies and premium receipts to confirm that we have taken out appropriate insurance under clause 6.3A of the contract.

[*List policies and receipts together with numbers and dates*]

[*Or*:]

We enclose documentary evidence that we maintain appropriate All Risks cover independently of our obligations under this contract.

[*Then add*:]

Please return all the enclosed papers by special or recorded delivery with a note stating that they are to the employer's satisfaction.

Yours faithfully

函件 24c

在接受标书或续保通知后21天内,致雇主
本函仅适用于 GC/Works/1(1998)合同
专递/挂号邮件

Letter 24c

To employer, within 21 days of acceptance of tender or renewal of insurance
This letter is only suitable for use with GC/Works/1 (1998)
Special/recorded delivery

尊敬的先生:

  根据合同条件第8(4)条规定,现附上我方保险人/保险经纪人[视情形取舍]开具的证实相应的保单已正式生效的证明。

你忠诚的

Dear Sir

In accordance with clause 8(4) of the conditions of contract, we have pleasure in enclosing a certificate from our insurer/broker [*delete as appropriate*] attesting that the appropriate insurance policies have been effected.

Yours faithfully

函件 24d

有关承包商保险,致雇主

本函仅适用于 GC/Works/1(1998)合同

专递/挂号邮件

Letter 24d

To employer, regarding contractor's insurance

This letter is only suitable for use with GC/Works/1 (1998)

Special/recorded delivery

尊敬的先生:

  你方[填入日期]的来函收悉。应你方要求,根据合同条件第 8 条规定,现附上下列保险单副本,请过目。

[根据序号和日期列出保险单清单]

你忠诚的

Dear Sir

Thank you for your letter of the [*insert date*]. In response to your request, we have pleasure in enclosing copies of the following insurance policies in accordance with clause 8 of the conditions of contract.

[*List policies together with numbers and dates*]

Yours faithfully

函件 25

保险人被认可后,致建筑师
本函不适用于 MW98 或 GC/Works/1(1998)合同
专递/挂号邮件

Letter 25

To architect, after insurers approved
This letter is not suitable for use with MW 98 or GC/Works/1 (1998)
Special/recorded delivery

尊敬的先生:

　　现附上已被认可的保险人[填入保险人名称]签发的号码为[填入号码]的保险单和于[填入日期]出具的保险费收据,根据本合同条件第 22A.2 条[使用 IFC98 合同时,为"第 6.3A.2 条"]规定,双方以共同的名义支付保险金。

你忠诚的

Dear Sir

We enclose insurance policy number [*insert number*] and premium receipt dated [*insert date*] in respect of insurance taken out with approved insurers, [*insert name*], for deposit with the employer in accordance with clause 22A.2 [*substitute* "6.3A.2" *when using IFC 98*] in joint names.

Yours faithfully

函件 26

有关雇主责任险,致建筑师

本函不适用于 MW98 或 GC/Works/1(1998)合同

专递/挂号邮件

Letter 26

To architect, regarding insurance of employer's liability

This letter is not suitable for use with MW 98 or GC/Works/1 (1998)

Special/recorded delivery

尊敬的先生:

你方 [填入日期] 签发的编号为 [填入编号] 的指令收悉,要求我方根据本合同条件第 22.2.1 条 [使用 IFC98 合同时,为"第 6.2.4 条"] 的规定,以双方共同的名义投保。我方建议向 [填入保险人名称] 投保,期盼得到雇主的认可,以便我方办理保险。

你忠诚的

Dear Sir

Thank you for your instruction number [*insert number*] dated [*insert date*] requiring us to take out and maintain joint names insurance under clause 21.2.1 [*substitute "6.2.4" when using IFC 98*]. We propose to take out the insurance with [*insert name*] and we should be pleased to receive the employer's approval to enable us to proceed.

Yours faithfully

函件 27
承担雇主责任险的保险人被认可后,致建筑师
本函不适用于 MW98 或 GC/Works/1(1998)合同
专递/挂号邮件

Letter 27
To architect, after approval of insurers for employer's liability
This letter is not suitable for use with MW 98 or GC/Works/1 (1998)
Special/recorded delivery

尊敬的先生:

　　现附上已被认可的保险人[填入保险人名称]于[填入日期]签发的编号为[填入号码]保险单和保险费收据。根据本合同条件第21.2.2条[使用IFC98合同时,为"第6.2.4条"]规定,双方应以共同的名义支付保险金。

你忠诚的

Dear Sir

We enclose insurance policy number [*insert number*] and premium receipt dated [*insert date*] in respect of insurance taken out with approved insurers, [*insert name*], for deposit with the employer in accordance with clause 21.2.2 [*substitute "6.2.4" when using IFC 98*] in joint names.

Yours faithfully

保险 41

函件 28
如需因承包商工期延误而给雇主造成的损失投保,致建筑师
本函不适用于 MW98 或 GC/Works/1(1998)合同

Letter 28

To architect, if insurance for employer's loss of liquidated damages is required
This letter is not suitable for use with MW 98 or GC/ Works/ 1（1998）

尊敬的先生：

你方于[填入日期]签发的编号为[填入编号]的指令收悉,现附上根据合同条件第 22D.1 条[使用 IFC98 合同时,为"第 6.3D.1 条"]规定,雇主要求投保的报价单。如有新的指示,请尽快告知。

你忠诚的

Dear Sir

Following receipt of your instruction number [*insert number*] dated [*insert date*] we have pleasure in enclosing a quotation for the insurance required by the employer under clause 22D.1 [*substitute "6.3D.1" when using IFC 98*] of the conditions of contract. Please let us have your further instructions as soon as possible.

Yours faithfully

函件 29

当承包商误期损害赔偿金的保险报价被接受时，致建筑师
本函不适用于 MW98 或 GC/Works/1(1998)合同
专递/挂号邮件

Letter 29

To architect, if quotation for liquidated damages insurance is to be accepted
This letter is not suitable for use with MW 98 or GC/Works/1 (1998)
Special/recorded delivery

尊敬的先生：

　　你方于[填入日期]签发的编号为[填入编号]的指令收悉。我方接受保险报价并办理了相关保险。现根据合同条件第22D.1条[使用IFC98合同时，为"第6.3D.1条"]规定，随函附上保险费收据，请查收。

你忠诚的

Dear Sir

Following receipt of your instruction number [*insert number*] of the [*insert date*] we have accepted the quotation and taken out the relevant policy. We enclose it herewith together with the premium receipt in accordance with the provisions of clause 22D.1 [*substitute "6.3D.1" when using IFC 98*] of the conditions of contract.

Yours faithfully

函件 30
当联合消防规范的补救措施作为变更事项,并要采取应急行动时,致建筑师
本函仅适用于 JCT98 合同

Letter 30
To architect, if Joint Fire Code remedial measures are a variation, but require emergency action
This letter is only suitable for use with JCT 98

尊敬的先生:

  根据合同条件第 22FC.3.1.2 条规定,现通知如下:我方已开始履行合同条件第 22FC.3.1 条规定的保险人所要求的与补救措施相一致的应急措施。具体步骤如下:

  [填入具体内容]

  我方在此确认,已完成的工程以及已提供的材料应作为变更事项处理。请提供进一步的指示。
  请将本函作为根据合同条件第 25.2.1.1 条规定下的工期延误通知以及根据合同条件第 26.1 条规定下的损失和(或)费用的索赔申请。

你忠诚的

Dear Sir

We hereby give notice as required by clause 22FC.3.1.2 that we have been obliged to commence the carrying out of work constituting emergency compliance with Remedial Measures as required by the insurers under the provisions of clause 22FC.3.1. The steps we are taking are as follows:

[*Insert details*]

We confirm that the work carried out and the materials supplied rank as a variation. Please provide further instructions necessary.
Please take this letter as a notice of delay under clause 25.2.1.1 and an application for loss and/or expense under clause 26.1.

Yours faithfully

函件 31a

有关雇主保险，致雇主
本函仅适用于 JCT98 合同和 WCD98 合同

Letter 31a

To employer, regarding employer's insurance
This letter is only suitable for use with JCT 98 and WCD 98

尊敬的先生：

　　我方期盼收到你方依据合同条件第 22B/22C 条 [视情形取舍] 的规定供我方审核的最后一次保险费收据。收据应表明该保险目前已生效，我方希望于 [填入日期] 之前收到该收据。

你忠诚的

Dear Sir

We should be pleased to receive for inspection your last premium receipt for insurance in accordance with clause 22B/22C [*delete as appropriate*] of the contract. The receipt should indicate that such insurance is currently effective and it should be in our hands by [*insert date*].

Yours faithfully

函件 31b
有关雇主保险,致雇主
本函仅适用于 IFC98 合同

Letter 31b
To employer, regarding employer's insurance
This letter is only suitable for use with IFC 98

尊敬的先生:

  我方期盼收到你方依据合同条件第 6.3B/6.3C 条 [视情形取舍] 规定供我方审核的最后一次保险费收据。收据应表明该保险目前已生效,我方希望于[填入日期]之前收到该收据。

你忠诚的

Dear Sir

We should be pleased to receive for inspection your last premium receipt for insurance in accordance with clause 6.3B/6.3C [*delete as appropriate*] of the contract. The receipt should indicate that the insurance is currently effective and it should be in our hands by [*insert date*].

Yours faithfully

函件 31c

有关雇主保险,致雇主

本函仅适用于 MW98 合同

Letter 31c

To employer, regarding employer's insurance

This letter is only suitable for use with MW 98

尊敬的先生:

  我方期盼收到你方依据合同条件第 6.3B 条规定供我方审核的最后一次保险费收据。我方根据合同条件第 6.4 条规定索要收据,以求证实该保险目前已生效。我方希望于[填入日期]之前收到该收据。

你忠诚的

Dear Sir

We should be pleased to receive for inspection your last premium receipt for insurance in accordance with clause 6.3B of the contract. We require the receipt under clause 6.4 in order to satisfy ourselves that the insurance is currently effective and it should be in our hands by [*insert date*].

Yours faithfully

函件 32a
当已投保的风险发生并造成损害时,致建筑师与雇主
本函仅适用于 JCT98 合同和 WCD98 合同

Letter 32a
To architect and employer, if any damage occurs due to an insured risk
This letter is only suitable for use with JCT 98 and WCD 98

尊敬的先生:

根据合同条件第 22A.4.1/22B.3.1/22C.4 条[视情形取舍]规定,现通知如下:正在施工的工程/场地材料[视情形取舍]已发生损失/损害[视情形取舍],该风险属于双方共同投保的风险之一。

[描述损失或损害情况,说明损失或损害的范围、性质和地点]

你忠诚的

Dear Sir

In accordance with clause 22A.4.1/22B.3.1/22C.4 [*delete as appropriate*] of the conditions of contract, we hereby give notice that loss/damage [*delete as appropriate*] has been caused to the work executed/site materials [*delete as appropriate*] by one of the risks covered by the joint names policy.

[*Describe the loss or damage, stating extent, nature and location*]

Yours faithfully

函件 32b
当已投保的风险发生并造成损害时,致建筑师与雇主
本函仅适用于 IFC98 合同

Letter 32b
To architect and employer, if any damage occurs due to an insured risk
This letter is only suitable for use with IFC 98

尊敬的先生:

　　根据合同条件第 6.3A.4.1/6.3B.3.1/6.3C.4 条 [视情形取舍] 规定,现通知如下:正在施工的工程/场地材料 [视情形取舍]已发生损失/损害[视情形取舍],该风险属于双方共同投保的风险之一。

　　[描述损失或损害情况,说明损失或损害的范围、性质和地点]

你忠诚的

Dear Sir

In accordance with clause 6.3A.4.1/6.3B.3.1/6.3C.4 [*delete as appropriate*] of the conditions of contract, we hereby give notice that loss/damage [*delete as appropriate*] has been caused to the work executed/site materials [*delete as appropriate*] by one of the risks covered by the joint names policy.

[*Describe the loss or damage, stating extent, nature and location*]

Yours faithfully

函件 33a
如果雇主未能使保险继续有效,致雇主
本函仅适用于 JCT98 合同和 WCD98 合同

Letter 33a
To employer, if he fails to maintain insurance cover
This letter is only suitable for use with JCT 98 and WCD 98

尊敬的先生:

　　今天你方与我方 [填入名字] 先生的电话谈话使我方确认,你方没有提供依据合同条件第 22B.2/22C.3 条[视情形取舍]规定要求的表明保险目前有效的证据。

　　考虑到该保险的重要性并且在无损于你方责任的前提下,我方将依据上述合同条件立即行使我方的权利。一旦你方提供了支付保险费的收据,我方有权将该款项加入合同总价中。

　　[如果未能根据合同条件第 22C.1 条规定投保,则添加以下内容:]

　　请注意,为了对保险事项进行调查并收集资料,我方计划大约在[填入日期][填入时间]行使权利进入现场并进行调查。

你忠诚的

抄送:建筑师

Dear Sir

We refer to your telephone conversation with our Mr [*insert name*] today and we confirm that you are unable to produce a receipt, as required by clause 22B. 2/22C. 3 [*delete as appropriate*], showing that insurance is currently effective.

In view of the importance of the insurance and without prejudice to your liabilities, we are arranging to exercise our rights under the above mentioned clause immediately. On production of a receipt for any premium paid, we will be entitled to have its amount added to the contract sum.

[*If a failure to insure under clause 22C. 1, add*:]

We draw your attention to the fact that we intend to exercise our right of entry and inspection for the purposes of survey and inventory on [*insert date*] at approximately [*insert time*].

Yours faithfully

Copy: Architect

函件 33b
如果雇主未能使保险继续有效，致雇主
本函仅适用于 IFC98 合同

Letter 33b
To employer, if he fails to maintain insurance cover
This letter is only suitable for use with IFC 98

尊敬的先生：

　　今天你方与我方[填入名字]先生的电话谈话使我方确认，你方未能提供依据合同条件第6.3B.2/6.3C.3条[视情形取舍]规定要求的表明保险目前有效的证据。

　　考虑到该保险的重要性并且在无损于你方责任的前提下，我方将依据上述合同条件立即行使我方的权利。一旦你方提供了支付保险费的收据，我方有权将该款项加入合同总价中。

[如果未能根据合同条件第6.3C.1条规定投保，则添加以下内容：]

　　请注意，为了对保险事项进行调查并收集资料，我方计划大约在[填入日期][填入时间]行使权利进入现场并进行调查。

你忠诚的

抄送：建筑师

Dear Sir

We refer to your telephone conversation with our Mr [*insert name*] today and we confirm that you are unable to produce a receipt, as required by clause 6.3B.2/6.3C.3 [*delete as appropriate*], showing that insurance is currently effective.

In view of the importance of the insurance and without prejudice to your liabilities, we are arranging to exercise our rights under the above mentioned clause immediately. On production of a receipt for any premium paid, we will be entitled to have its amount added to the contract sum.

[*If a failure to insure under clause 6.3C.1, add*:]

We draw your attention to the fact that we intend to exercise our right of entry and inspection for the purposes of survey and inventory on [*insert date*] at approximately [*insert time*].

Yours faithfully

Copy: Architect

函件 33c
如果雇主未能使保险继续有效,致雇主
本函仅适用于 MW98 合同

Letter 33c
To employer, if he fails to maintain insurance cover
This letter is only suitable for use with MW 98

尊敬的先生:

今天你方与我方[填入名字]先生的电话谈话使我方确认,你方未能提供依据合同条件第 6.4 条要求的保险证明材料来证实合同条件第 6.3B 条规定的保险已办理并且在规定的时间内有效。

你方违反了合同,就此我方准备对由此产生的损害进行索赔。我方认为该损害赔偿金应包括我方办理相应保险时所发生的费用。

你忠诚的

Dear Sir

We refer to your telephone conversation with our Mr [*insert name*] today and we confirm that you are unable to produce evidence, as required by clause 6.4, that the insurance referred to in clause 6.3B has been taken out and is in force at all material times.

You are in breach of contract for which we intend to claim damages. We are advised that such damages include our costs in taking out appropriate insurance.

Yours faithfully

第四章　施工现场活动

除非项目很小，在施工阶段中将要撰写很多信函。一些相当规模的事务，例如付款、工期延长、损失和（或）费用，为便于寻找，已在本书的其他章节中涉及。但仍有很多其他事务是属于本章标准函件范围的。

撰写重要信函的内容可能涉及任命工地代理人、经理或工长、总进度、现场占有、图纸、现场会议、指令、缺陷材料和文物。没有发现 WCD98 合同涵生的信函与传统合同的信函有什么关系。图纸的提交及评审，业主的要求和承包商的建议本身或它们之间的差异，法规的变化或开发控制要求的变化等等，这些主题都将包括在本章内。

有些时候，承包商与工程管理人员发生一些问题，经常不知道下一步该怎么做，因为这些管理人员常有建筑师的支持。没有理由让承包商容忍一些不正规的做法，因此本章内一部分信函表达经常发生的令人遗憾的情形。

无可置疑，在此阶段写就的大部分信函都不一定是标准的格式。撰写这些信函应反映各种具体场合。掌握日常事务的标准信函格式会在处理那类非常见场合时可省一些时间。

函件 34a
致建筑师,有关负责人或代理人事项

Letter 34a
To architect, regarding person-in-charge or agent

尊敬的先生:

　　谨通知您,本项目现场负责人[当使用 GC/Works/1(1998)合同时,代之以"代理人"]为[填入姓名]。他/她[视情况取舍]是有能力并有此类工程的施工经验。

你忠诚的

Dear Sir

This is to inform you that the person-in-charge [*substitute "agent" when using GC/ Works/ 1 (1998)*] of this project on site will be [*insert name*]. He/She [*delete as appropriate*] is competent and experienced in this type of work.

Yours faithfully

函件 34b

致建筑师,有关现场经理的任命

本信仅适用于 WCD98 合同

Letter 34b

To architect, regarding the appointment of a site manager
This letter is only suitable for use with WCD 98

尊敬的先生:

  我方提议任命[填入姓名]作为本项目的现场经理。他/她[视情况取舍]有能力并具有此类工程的施工经验,如能收到你方根据合同补充条款 S3.2 而签发的对此任命的书面批准,我方将十分高兴。

你忠诚的

Dear Sir

We propose to appoint [*insert name*] as site manager for this project. He/She [*delete as appropriate*] is competent and experienced in this type of work and we should be pleased to receive your written consent to this appointment as required by supplementary provision S3.2 of the contract.

Yours faithfully

函件 35
致建筑师,有关批准现场经理的撤免或更换
本函件仅适用于 WCD98 合同

Letter 35
To architect, regarding consent to removal or replacement of site manager
This letter is only suitable for use with WCD 98

尊敬的先生:

[若撤免]

依据[填入日期]的电话商谈,基于双方商讨的理由,我方在此确认我方要撤免现场经理[填入姓名]。我方知道你方接受这些理由,从[填入日期]起将任命[填入姓名]为现场经理。依据合同补充条款 S3.2 的要求,我方希望得到你方的书面批准。

[若更换]

现在的现场经理[填入姓名],将于[填入日期]离任。从该日期起将任命[填入姓名]为现场经理。我方将高兴地收到你方依据合同补充条款 S3.2 而签发的对此任命的书面批准。

你忠诚的

Dear Sir

[*If removal*:]

We refer to our telephone conversation of the [*insert date*] and confirm that we intend to remove [*insert name*] as site manager for the reasons we discussed. We understand that you accept those reasons and we should be pleased to receive your written consent, as required by supplementary provision S3.2 of the contract, to the appointment of [*insert name*] as site manager from [*insert date*].

[*If replacement*:]

The present site manager, [*insert name*], will be leaving on [*insert date*] and we should be pleased to receive your written consent, as required by supplementary provision S3.2 of the contract, to the appointment of [*insert name*] as site manager from that date.

Yours faithfully

函件 36
若要求提供现场人员的姓名及地址,致建筑师
本函仅适用于 GC/Works/1(1998)合同

Letter 36

To architect, if required to furnish names and addresses of operatives
This letter is only suitable for use with GC/Works/1 (1998)

尊敬的先生:

　　你方[填入日期]的来函收悉,谢谢。
　　依据合同条件第26(2)条,我方高兴地附上与工程有关的那些人员的姓名、地址以及他们所从事工作的能力状况。若要进一步的细节,请通知我方。他们都是我方雇用的人员,我方对他们负有责任。
　　施工人员中,有些可能不是我方的雇员,我方无法提供这些人员的资料,而且我方也无权得到这些资料。

你忠诚的

Dear Sir

Thank you for your letter of the [*insert date*].

In accordance with clause 26(2) of the conditions of contract, we have pleasure in enclosing a full list of the names and addresses of persons who are concerned with the works and the capacities in which they are so concerned. Please inform us if you require further particulars. They are all persons whom we employ and for whom we are responsible.

There may well be other persons not in our employ who are concerned with the works for whom we do not have, nor do we have the authority to obtain, such information.

Yours faithfully

函件 37

若需要出入证件,致建筑师
本函仅适用于 GC/Works/1(1998)合同

Letter 37

To architect, if passes are required
This letter is only suitable for use with GC/Works/1 (1998)

尊敬的先生:

  依据合同条件第 27 条,我方在此附上一份人员名单,请为他们进入工地现场提供出入证件。若还需要这些人员的其它详细资料,请通知我方。

你忠诚的

Dear Sir

In accordance with clause 27 of the conditions of contract, we enclose a list of persons requiring passes to secure admission to the site. Please inform us what, if any, other details you require in respect of each person.

Yours faithfully

函件 38a
若在合同规定日或延期日未能提供工地现场占有权,致雇主
本函不适用于 MW98 合同或 GC/Works/1(1998)合同
专递/挂号邮件

Letter 38a
To employer, if possession not given on the due or the deferred date
This letter is not suitable for use with MW 98 or GC/Works/1 (1998)
Special/recorded delivery

尊敬的先生:

[若没有延期占有工地的条款]

依据合同条件第23.1.1条[当使用 IFC98 合同时,为"第2.1条"]工地现场应在[填入日期]移交给我方。但在该日期并没有向我方移交工地现场。

[若延期占有工地条款适用,但没有遵守延期占有日]

依据第23.1条[当使用 IFC98 合同时,为"第2.2条"]你方把工地占有日推延至[填入日期]。但在该延期日你方仍未让我方接收工地现场。

[然后,任选下述一段]

况且,建筑师已于今日电话通知我方,他不能确认你方将何时向我方移交工地现场。

[或]

建筑师今日电话通知在[填入日期]之前你方将不能移交工地现场。

[然后]

这是一种严重的违约行为,对此我方要求获得适当的补偿,我方保留为此事要求赔偿的一切权利。在不影响前述权利情况下,我方认为召开一次会议是有益的,期盼得到你方的回复。

你忠诚的

Dear Sir

[*If deferment of possession clause does not apply*:]

Possession of the site should have been given to us on [*insert date*] in accordance with clause 23. 1. 1 [*substitute* "2. 1" *when using IFC 98*] of the conditions of contract. Possession was not given to us on the due date.

[*If deferment of possession clause does apply, but deferred date not met*:]

Under clause 23. 1. 2 [*substitute clause* "2. 2" *when using IFC 98*] you deferred the giving of possession until [*insert date*]. You did not give us possession on the deferred date.

[*Then either*:]

Moreover, the architect has informed us today by telephone that he is unable to say with certainty when you will be able to give possession.

[*Or:*]

The architect has informed us today by telephone that you will be unable to give possession until [*insert date*].

[*Then*:]

This is a serious breach of contract for which we will require appropriate compensation and we reserve all our rights and remedies in this matter. Without prejudice to the foregoing, we suggest that a meeting would be useful and look forward to hearing from you.

Yours faithfully

函件 38b

若在规定日期没有移交工地，致雇主

本函仅适用于 MW98 合同

专递/挂号邮件

Letter 38b

To employer, if possession not given on the due date

This letter is only suitable for use with MW 98

Special / recorded delivery

尊敬的先生：

依据合同条件第 2.1 条，移交了工地才能使我方于[填入日期]开工。但在该日期没有向我方移交工地。

[然后任选下述一段]

建筑师已于今日电话通知我方，他不能确认你方将何时能移交工地。

[或]

建筑师已于今日电话通知我方，在[填入日期]之前你方将不能移交工地。

[然后]

这是一种严重的违约行为，对此事件，我方要求得到适当的补偿，我方保留为此要求赔偿的一切权利。在不影响前述权利情况下，我方认为召开一次会议是有益的，希望得到你方的回复。

你忠诚的

Dear Sir

Possession of the site should have been given to enable us to commence the works on [*insert date*] in accordance with clause 2.1 of the conditions of contract. Possession was not given to us on the due date.

[*Then either*:]

The architect has informed us today by telephone that he is unable to say with certainty when you will be able to give possession.

[*Or*:]

The architect has informed us today by telephone that you will be unable to give possession until [*insert date*].

[*Then*:]

This is a serious breach of contract for which we will require appropriate compensation and we reserve all our rights and remedies in this matter. Without prejudice to the foregoing, we suggest that a meeting would be useful and look forward to hearing from you.

Yours faithfully

函件 38c 若得到通知在规定日期不能移交工地,而且延期条款不适用时,致雇主 本函不适用于 GC/Works/1(1998)合同 专递/挂号邮件 **Letter 38c** To employer, if notice received that possession cannot be given on the due date and the deferment clause does not apply *This letter is not suitable for use with GC/Works/1 (1998)* *Special/recorded delivery*

尊敬的先生:

  我方已从建筑师处得到通知,你方不能在[填入日期]向我方移交工地现场。

  这是对合同条件第23.1.1条[当使用IFC98合同或MW98合同时,为"第2.1条"]的严重违反,对此,我方要求得到适当补偿,并且我方保留为此要求赔偿的一切权利。

  在不影响前述权利情况下,我方要求召开一次紧急会议讨论此问题,希望得到你方的答复。

你忠诚的

Dear Sir

We have received notice from the architect that you will be unable to give us possession of the site on [*insert date*].

This will be a serious breach of clause 23.1.1 [*substitute "2.1" when using IFC 98 or MW 98*] of the conditions of contract for which we will require appropriate compensation and we reserve all our rights and remedies in this matter.

Without prejudice to the foregoing, we request an urgent meeting to discuss the problem and we look forward to hearing from you.

Yours faithfully

施工现场活动　　65

函件 39
若工地占有日提前时,致建筑师
本函不适用于 GC/Works/1(1998)合同

Letter 39
To architect, if date for possession advanced
This letter is not suitable for use with GC/Works/1(1998)

尊敬的先生:

　　你方于[填入日期]来函收悉,谢谢。来函告知雇主允许我方在[填入日期]接收现场。

　　当然,我方将尽力利用这一有利机会,比预期提早开工。我方将通知你方何时接收工地。因为有很多事项须考虑,劳动力资源要重新组织,这一切都将产生额外成本,我方期盼雇主能对这些给予补偿。若雇主将对这些费用作出补偿,请尽快告知我方,以便我方能作出相应的决定。若我方在你方提议的日期接收工地,这一行为将不影响我方在附件中规定的在竣工日完成工程的施工。

你忠诚的

Dear Sir

Thank you for your letter of the [*insert date*] advising us that the employer is prepared to allow us to take possession of the site on [*insert date*].

Naturally, we will endeavour to take advantage of the opportunity to make an earlier start than we anticipated and we will inform you when we intend to take possession. There are many things to take into consideration and some reorganising of our labour resources would be involved, all of which would result in additional costs which we would expect the employer to reimburse. Please let us know, by return, if the employer will reimburse such costs so that we can make our decision accordingly. If we take possession of the site on the date you suggest, such action will not affect our obligation to complete the works on the date for completion stated in the appendix.

Yours faithfully

函件 40

致雇主,同意聘用其他人员
本函不适用于 MW98 合同或 GC/Works/1(1998)合同

Letter 40

To employer, giving consent to the engagement of other persons
This letter is not suitable for use with MW 98 or GC/Works/1(1998)

尊敬的先生:

　　我方已收到你方[填入日期]的来函/建筑师的指令[视情况取舍]。关于根据合同条件第29.2条[当使用IFC98合同时,为"第3.11条"],同意对[填入姓名]的聘用。该项工作在合同清单[或技术规范]中未具体列出,但我方知道该工作应包括[填入详细内容],并将于[填入日期]在现场开工。

　　若你方能书面确认同意我方的工期延长及损失和(或)费用补偿的要求,我方将同意实施此项工作。

你忠诚的

抄送:建筑师

Dear Sir

We are in receipt of your letter/an architect's instruction [*delete as appropriate and insert date*] referring to the engagement of [*insert name*] in accordance with clause 29.2 [*substitute "3.11" when using IFC 98*] of the conditions of contract. The work is not detailed in the contract bills [*or specification*], but we understand that it will consist of [*insert details*] and it will commence on site on the [*insert date*].

We give our consent to the work if you will confirm in writing that you recognise our entitlement to extension of time and loss and/or expense in consequence.

Yours faithfully

Copy: Architect

函件 41

致雇主，拒绝聘用其他人员

本函不适用于 MW98 合同或 GC/Works/1(1998)合同

Letter 41

To the employer, withholding consent to the engagement of other persons

This letter is not suitable for use with MW 98 or GC/Works/1 (1998)

尊敬的先生：

　　我方已收到你方[填入日期]的来函/建筑师的指令[视情况取舍]，从中我方知道，依据合同条件第 29.2 条[当使用 IFC98 合同时，为"第 3.11 条"]，同意聘用[填入姓名]。该项工作不曾在合同清单[或技术规范]中具体开列，但我方知道它将包括[填入详细内容]，并要求于[填入日期]在工地上开工，并于[填入日期]竣工。

　　对此工作我方不能给予同意，因为[填入合理的反对理由]。

你忠诚的

抄送：建筑师

Dear Sir

We are in receipt of your letter/an architect's instruction [*delete as appropriate and insert date*] from which we see that it is intended to engage [*insert name*] in accordance with clause 29.2 [*substitute "3.11" when using IFC 98*] of the conditions of contract. The work is not detailed in the contract bills [*or specification*], but we understand that it will consist of [*insert details*] and it is proposed to commence on site on [*insert date*] reaching completion on the [*insert date*].

We are unable to give our consent to the work, because [*insert details of reasonable objection*].

Yours faithfully

Copy: Architect

函件 42
有关现场会议纪要的内容,致建筑师

Letter 42
To architect, regarding items in minutes of site meeting

尊敬的先生:

我方审查了今日收到的[填入日期]的会议纪要。我方提出下列意见:

[列出意见]

请把上述意见列入会议纪要并在下次会议时公布。

你忠诚的

抄送:所有与会者及原会议纪要抄送人员

Dear Sir

We have examined the minutes of the meeting held on the [*insert date*] which we received today. We have the following comments to make:

[*List comments*]

Please arrange to have these comments published at the next meeting and inserted in the appropriate place in the minutes.

Yours faithfully

Copies: All present at meeting and those included in the original circulation

函件 43a

致建筑师,提交总进度计划

本函仅适用于 JCT98 合同

Letter 43a

To architect, enclosing master programme

This letter is only suitable for use with JCT 98

尊敬的先生:

依据合同条件第 5.3.1.2 条,我十分高兴地送上我方的总进度计划二份。我方期待能尽快地收到你方的批准,以便我方能按计划实施项目。

你忠诚的

Dear Sir

In accordance with clause 5.3.1.2 of the conditions of contract, we have pleasure in enclosing two copies of our master programme.

We should be pleased to receive your approval as soon as possible so that we can proceed with project planning.

Yours faithfully

函件 43b
致建筑师，提交总进度计划
本函不适用于 JCT98 合同

Letter 43b
To architect, enclosing master programme
This letter is not suitable for use with JCT 98

尊敬的先生：

　　十分高兴地送上我方的总进度计划两份。请尽快给予批准，以便我方能按计划实施项目。

你忠诚的

Dear Sir

We have pleasure in enclosing two copies of our master programme. Please let us have your approval as soon as possible so that we can proceed with project planning.

Yours faithfully

函件 44
致建筑师,提交总进度计划的修改稿
本函仅适用于 JCT98 合同

Letter 44
To architect, enclosing revision to the master programme
This letter is only suitable for use with JCT 98

尊敬的先生:

    依据合同条件第 5.3.1.2 条,十分高兴地送上我方的总进度计划第[填入编号]号修改稿二份,它考虑了你方依据第 25.3.1 条/33.1.3 条[视情况取舍]作出的决定。
    我方将高兴地期待获得你方对该修改进度计划的批准。

你忠诚的

Dear Sir

In accordance with clause 5.3.1.2 of the conditions of contract, we have pleasure in enclosing two copies of our master programme revision [*insert revision number*] to take account of your decision under clause 25.3.1/33.1.3 [*delete as appropriate*].

We should be pleased to have your approval to the revised programme.

Yours faithfully

函件 45
致建筑师，提交承包商的结算单
本函仅适用于 JCT98 合同

Letter 45
To architect, sending Contractor's Statement
This letter is only suitable for use with JCT 98

尊敬的先生：

　　[任选下述一段]

　　我方参阅了[填入编号]图纸及涉及实施规定工程的合同清单的第[填入编号]号工作细目。

　　[或]

　　我方参阅了你方于[填入日期]签发的有关实施规定工程的暂定金额的第[填入编号]号指令。

　　[续写]

　　依据合同条件第 42.2 条，在此送上我们承包商的结算单，它在格式和细节上都详细地对我方的要求作出了解释。我方希望能马上收到你方的批准，以便我方毫无迟延地实施工程。

你忠诚的

Dear Sir

[*Either*:]

We refer to drawings numbered [*insert numbers*] and to item number [*insert number*] of the contract bills which refer to performance specified work.

[*Or*:]

We refer to your instruction number [*insert number*] dated [*insert date*] on the expenditure of a provisional sum for performance specified work.

[*Then*:]

In accordance with clause 42.2 of the conditions of contract, we enclose our Contractor's Statement which is sufficient in form and detail adequately to explain our proposals. We should be pleased to receive your approval by return in order to allow us to proceed with the works without delay.

Yours faithfully

函件 46
若指令要求实施附件中未列明的工程,致建筑师
本函仅适用于 JCT98 合同

Letter 46
To architect, if instruction requires work not identified in the appendix
This letter is only suitable for use with JCT 98

尊敬的先生:

收到你方于[填入日期]发出的第[填入号码]号指令,要求我方实施规定的工程。但附件中没有列明这些工程,请参阅合同条件第42.12条。

[然后加入]

我方不准备实施此项工程。

[或]

我方在满足下列条件情况下,可以同意实施此项工程。

[列出条件]

你忠诚的

Dear Sir

We are in receipt of your instruction number [*insert number*] dated [*insert date*] which purports to instruct us to carry out performance specified work. Such work has not been identified in the appendix and we refer you to clause 42.12 of the conditions of contract.

[*Then add*:]
We are not prepared to carry out this work.

[*Or:*]
We may be prepared to agree to carry out this work under the following conditions:

[*List conditions*]
Yours faithfully

施工现场活动   75

函件 47
若建筑师的指令严重影响规定工程的实施,致建筑师
本函仅适用于 JCT98 合同

Letter 47
To architect, if instruction injuriously affects performance specified work
This letter is only suitable for use with JCT 98

尊敬的先生:

收到你方于[填入日期,该日期与本函日期之间的时间不应超过 7 天]发出的要求我方实施[填入指令内容]的第[填入编码]号指令。

请把本函作为依据合同条件第 42.15 条的书面通知,我方认为该指令会严重影响规定工程实施的效率,因为[填入原因]。

[然后,任选下述一段]

若你方能修正指令以避免带来不利影响,我方将不胜感激。

[或]

我方将在下列条件下准备实施你方的指令:

[列出条件]

你忠诚的

Dear Sir

We are in receipt of your instruction number [*insert number*] dated [*insert date which should not be more than 7 days before the date of the letter*] which instructs us to [*insert the substance of the instruction*].

Please take this as notice in writing in accordance with clause 42.15 of the conditions of contract that, in our opinion, the instruction injuriously affects the efficacy of the performance specified work, because [*insert reasons*].

[*Then either*:]

We should be pleased if you would amend your instruction to remove the injurious affection.

[*Or*:]

We will be prepared to carry out your instruction on the following conditions:

[*List conditions*]

Yours faithfully

函件 48
致建筑师,请求提供资料

Letter 48
To architect, requesting information

尊敬的先生:

我方希望能得到下列资料。为了能依据合同条件,实施和完成该工程,我方必须在规定的日期之前得到这些资料,这是很必要的。

[列出要求的资料及必须收到每项资料的日期,凡有可能,允许建筑师至少有14天时间准备资料]

你忠诚的

Dear Sir

We should be pleased to receive the following information which it is necessary for us to receive by the dates stated in order to enable us to carry out and complete the works in accordance with the conditions of contract.

[*List the information required and the date by which each item of information must be received. Wherever possible, allow the architect at least 14 days to prepare the information*]

Yours faithfully

函件 49

若图纸上有关放线资料不全,致建筑师

本函不适用于 WCD98 合同

Letter 49

To architect, if insufficient information on setting out drawings

This letter is not suitable for use with WCD 98

尊敬的先生:

  我方计划于[填入日期]在工地开工。我方的首要工作是工程放线。细查你方提供的图纸后,发现资料不全,以致我方不能正确放线。现附上你方编号为[填入图号]图纸的拷贝件,我方已在图中用红笔勾出需要尺寸/标高的位置[视情况取舍]。

  为了避免工程延误和中断,我方要求在[填入日期]之前得到这些资料。

你忠诚的

Dear Sir

We are preparing to commence work on site on the [*insert date*]. Our first task will be to set out the works. An examination of the drawings you have supplied to us reveals that there is insufficient information for us to set out the works accurately. We enclose a copy of your drawing number [*insert number*] on which we have indicated in red the positions where we need dimensions/levels [*delete as appropriate*].

We need this information by [*insert date*] in order to avoid delay and disruption to the works.

Yours faithfully

函件 50
致建筑师,要求确认放线定位无误
本函不适用于 WCD98 合同

Letter 50
To architect, requesting information that setting out is correct
This letter is not suitable for use with WCD 98

尊敬的先生:

依据我方[填入日期]的函件,告知你方所给的图纸上资料不全,使我方不能正确放线。你方电话答复要我方依据提供的这些资料尽我方的最大努力做好工程放线。我方已执行了你方的指令,希望得到你方确认放线定位正确的答复。如果我方不能立即得到你方邮寄的书面确认,我方不得不通知你方,工期将要延误和中断,为此,我方将寻求合理的经济补偿。

你忠诚的

Dear Sir

We refer to our letter of the [*insert date*] in which we informed you that the information on your drawings was insufficient to enable us to set out the works accurately. You responded by telephone, asking us to set out the works to the best of our ability based on the information provided. We have carried out your instructions and we should be pleased to receive your confirmation that the setting out is correct. If we do not receive such confirmation, in writing, by return of post, we shall be obliged to notify you of delay to the works and disruption for which we will seek appropriate financial recompense.

Yours faithfully

函件 51
当迟收到资料时,致建筑师
本函仅适用于 JCT98 合同及 IFC98 合同

Letter 51
To architect, if information received late
This letter is only suitable for use with JCT 98 and IFC 98

尊敬的先生:

　　[填入日期]我方要求在[填入日期]前得到下列图纸/详图/指令[视情况取舍]。现仍未收到。

　　合同条件第5.4.2条[使用IFC98合同时,为"第1.7.2条"]要求你方在"必要时及时"提供这些资料,你方明显地未履行此义务。我方履行了该条最后部分规定的我方义务,并提早通知你方我方必须获得这些资料的时间。资料的欠缺正在造成我方的工期延误和工程中断。请即刻告知我方何时能得到这些资料。

　　请把本函作为我方依据第25.2.1.1条[当使用IFC98合同时,为"第2.3条"]发出的延误通告和依据第26.1条[当使用IFC98合同时,为"第4.11条"]发出要求对损失和(或)费用补偿的申请。我方期盼能在今后适当时间提交进一步的细节。

你忠诚的

Dear Sir

On [*insert date*] we requested the following drawings/details/instructions [*delete as appropriate*] by [*insert date*]. They have not been received.

Clause 5.4.2 [*substitute* "1.7.2" *when using IFC 98*] requires you to provide such information 'as and when from time to time may be necessary'. You are clearly in breach of your obligations. We have complied with our duty in the last part of the clause, to advise you sufficiently in advance of the time when it is necessary for us to receive the information. The absence of the information is causing us delay and disruption. Please inform us, by return, when we can expect it to arrive.

Take this as notice of delay under clause 25.2.1.1 [*substitute* "2.3" *when using IFC 98*] and an application for loss and/or expense under clause 26.1 [*substitute* "4.11" *when using IFC 98*]. We expect to be able to provide further details in due course.

Yours faithfully

施工现场活动　　81

函件 52
若按资料提交时间表未收到资料时,致建筑师
本函仅适用于 JCT98 合同及 IFC98 合同

Letter 52
To architect, if information not received in accordance with the information release schedule
This letter is only suitable for use with JCT 98 and IFC 98

尊敬的先生:

　　我方提醒你方注意本项目的资料提交时间表,它规定[描述资料内容及应提交的日期]。但该资料至今尚未收到。

　　很明显,你方未履行第 5.4.1 条[当使用 IFC98 合同时,为"第 1.7.1 条"]规定的义务。由于未提供资料,正在造成我方工程延误和工程中断。收到本函,请尽快告知我方何时能得到这些资料。

　　请把本函作为我方依据合同条款第 25.2.1.1 条[当使用 IFC98 合同时,为"第 2.3 条"]发出的延误通告和依据第 26.1 条[当使用 IFC98 合同时,为"第 4.11 条"]发出要求对损失和(或)费用补偿的申请。我方期盼能在今后适当时间提交进一步的细节。

你忠诚的

Dear Sir

We draw your attention to the information release schedule for this project which indicates that [*describe the information and the date on which it should have been provided*]. It has not been received.

You are clearly in breach of your obligations as set out in clause 5.4.1 [*substitute "1.7.1" when using IFC 98*]. The absence of the information is causing us delay and disruption. Please inform us, by return, when we can expect it to arrive.

Take this as notice of delay under clause 25.2.1.1 [*substitute "2.3" when using IFC 98*] and an application for loss and/or expense under clause 26.1 [*substitute "4.11" when using IFC 98*]. We expect to be able to provide further details in due course.

Yours faithfully

函件 53
当图纸上发现设计错误时,致建筑师
本函不适用于 WCD98 合同

Letter 53
To architect, if design fault in drawings
This letter is not suitable for use with WCD 98

尊敬的先生:

  我方希望你方能对你方提供的[填入图号]图纸给以进一步的关注。依我方的观点[描述问题]。因此,若我方按此未经修改的图纸继续施工,它将会/可能[视情况取舍]带来严重的缺陷或损失,我方对此不负责任。

  请告知你方的意见并在[填入日期]前提交修改后的图纸,以避免工程进度的延误和中断。

你忠诚的

Dear Sir

We should be pleased if you would give further consideration to your drawing number [*insert number*]. In our view [*describe the problem*]. Therefore, if we proceed on the basis of this drawing without amendment, it is likely/possible [*delete as appropriate*] that serious defects or damage will result for which we can accept no liability.

Please let us have your comments and amended drawing by [*insert date*] to avoid delay and disruption to the progress of the works.

Yours faithfully

函件 54

若建筑师拒绝修改设计错误时,致建筑师

本函不适用于 WCD98 合同

专递/挂号邮件

Letter 54

To architect, if a design fault which he refuses to correct

This letter is not suitable for use with WCD 98

Special/ recorded delivery

尊敬的先生:

　　我方于[填入日期]的函中提醒你方,在编码为[填入图号]的图纸中存在设计错误。我方注意到,你方[填入日期]的来函表示拒绝考虑我方要求修改设计的意见。

　　我方就已知的设计错误告诫了你方及雇主,免除了我方的责任。请把本函作为正式通知,我方将严格按照上述图纸,执行你方的指令,继续施工。但我方对此造成的差错将不承担任何责任。

你忠诚的

抄送:雇主

[注:若设计错误涉及健康或安全,本函将不恰当]

Dear Sir

We refer to our letter of the [*insert date*] warning you of a design fault in your drawing number [*insert number*]. We note that, by your letter of the [*insert date*], you decline to amend your design to take account of our comments.

We have discharged any duty we may have to warn you and the employer of known design defects. Please take this as formal notice that we will carry out your instructions to proceed to construct in strict accordance with the above mentioned drawing, but we will not accept any liability for subsequent failure.

Yours faithfully

Copy: Employer

[*Note*: *This letter is unlikely to be adequate if the design failure will endanger health or safety*]

函件 55
若承包商提供图纸时,致建筑师
本函仅适用于 WCD98 合同

Letter 55
To architect, if contractor providing drawings
This letter is only suitable for use with WCD 98

尊敬的先生:

依据合同条件第 5.3 条,现附上两份下列文件:[列出图纸,技术规范,详图,标高及放线尺寸]

以上文件仅供参阅

你忠诚的

Dear Sir

In accordance with clause 5.3 of the conditions of contract, we enclose two copies of the following:

[*List drawings, specifications, details, levels and setting out dimensions*]

These documents are provided for information only.

Yours faithfully

函件 56
当他退回承包商的图纸并提出意见时,致建筑师
本函仅适用于 WCD98 合同

Letter 56
To architect, if he returns contractor's drawings with comments
This letter is only suitable for use with WCD 98

尊敬的先生:

你方的[填入日期]的来函及退回的[填入图号]图纸和意见收到,谢谢。

尽管我方很感激你方提出的意见,但我方必须指出,合同条件第 5.3 条的目的是向你方提交我方计划在工程中使用的图纸及其他资料。这些图纸是严格按照"承包商方案"而设计的,它体现了"雇主要求"。

因此,若要求我方执行你方的意见,请依据第 12.2.1 条的规定,把它们作为变更指令签发。若我方不反对,这样的指令将依据第 12.4 条/12.5 条/补充规定 S6[视情况取舍]来评估,工期延期将依据第 25 条来测算,损失和(或)费用将依据第 26 条/补充规定 S7[视情况取舍]来计价。

你忠诚的

Dear Sir
Thank you for your letter of the [*insert date*] with which you returned our drawings number [*insert numbers*] with comments.

Although we are always grateful for any comments you feel able to make, we should point out that the purpose of clause 5.3 of the conditions of contract is to provide you with copies of the drawings and other information we intend to use to carry out the work. The drawings have been prepared strictly in accordance with our Contractor's Proposals which are a response to the Employer's Requirements.

If, therefore, you intend us to act upon your comments, please issue them as change instructions under the provisions of clause 12.2.1. If we have no reasonable objection, such an instruction would be subject to valuation under clause 12.4/12.5/supplementary provision S6 [*delete as appropriate*], extension of time under clause 25 and loss and/or expense under clause 26/supplementary provision S7 [*delete as appropriate*].

Yours faithfully

函件 57
当承包商提交图纸听取意见时,致建筑师
本函仅适用于 WCD98 合同

Letter 57
To architect, if contractor submitting drawings for comment
This letter is only suitable for use with WCD 98

尊敬的先生:

依据合同补充条 S2 及"雇主要求"中的第[填入条款号]条,我方附上下列文件二份:

[列出图纸,详图,文件或其他资料]

为了避免工期的延误,若能在本函日期起的 5 个工作日[或按照"雇主要求"中规定的天数]内收到你方的意见或批准,我方将不胜感激。

你忠诚的

Dear Sir

In accordance with supplementary provision S2 of the contract and clause [*insert clause number*] of the Employer's Requirements, we enclose two copies of the following documents:

[*List drawings, details, documents or other information*]

We should be pleased to receive your comments or approval within five [*or such number as is stated in the Employer's Requirements*] working days of the date of this letter in order to avoid delay to the project.

Yours faithfully

函件 58
当建筑师退回承包商的图纸并提有意见时,致建筑师
本函仅适用于 WCD98 合同

Letter 58
To architect, if he returns contractor's drawings with comments
This letter is only suitable for use with WCD 98

尊敬的先生:

收到你方于[填入日期]的来函及附有你方意见的图号为[填入图号]的图纸各份一份,谢谢。

依据补充条款 S2,我方考虑了你方的意见,现随函附上已作适当修改的图纸二份。修改后的图号为[填入图号]。

为了避免进一步延误工期,若能在本函日期起的 5 个工作日内收到你方的批准或意见,我方将不胜感激。

你忠诚的

Dear Sir

Thank you for your letter of the [*insert date*] with which you enclosed one copy of each of our drawings numbered [*insert numbers*] with your comments.

In accordance with supplementary provision S2, we have taken account of your comments and we herewith submit two further copies of such drawings suitably revised. The revised numbers are [*insert numbers*].

We should be pleased to receive your approval or comments within 5 working days of the date of this letter in order to avoid further delay.

Yours faithfully

函件 59

若建筑师退回承包商的图纸并提出不合理的意见时,致建筑师

本函仅适用于 WCD98 合同

Letter 59

To architect, if he returns contractor's drawings with unreasonable comments

This letter is only suitable for use with WCD 98

尊敬的先生:

你方[填入日期]的来函及附有你方意见的图号为[填入图号]各一份图纸已收到。

我方提交的图纸是严格遵照"承包商方案"的,它体现了"雇主要求"。你方并未表示不同的意见。就它们的原意而言,甚至可以认为,它们是依据第 12.2.1 条的规定而签发的变更令。

因此,你方的意见应撤销,若你方愿意,请把这些意见作为变更指令签发。请从本函日期起的 5 个工作日内处理好此事,否则我方将蒙受工期延误和工程中断,如此,我方将依据合同的有关条款要求工期延长和经济补偿。

你忠诚的

Dear Sir

We are in receipt of your letter of the [*insert date*] with which you enclosed one copy of each of our drawings numbered [*insert numbers*] with your comments.

Our submitted drawings are strictly in accordance with our Contractor's Proposals in response to the Employer's Requirements. Your comments do not suggest the contrary, rather, on their true interpretation, they are change instructions which should be issued under the provisions of clause 12.2.1.

You should, therefore, withdraw your comments and, if you so wish, issue them as a change instruction. Please act within 5 working days from the date of this letter or we shall suffer delay and disruption for which we shall require an extension of time and financial reimbursement under the appropriate clauses of the contract.

Yours faithfully

函件 60
若建筑师不能在规定的时限内退回承包商的图纸时,致建筑师
本函仅适用于 WCD98 合同

Letter 60
To architect, if he fails to return the contractor's drawings in due time
This letter is only suitable for use with WCD 98

尊敬的先生:

在此谨想提及我方[填入日期]的函及附上的二份请求批准和征求意见的图号为[填入图号]图纸。这些意见或批准应在本函日期起的5个工作日内[或按"雇主要求"中规定的天数]提供。但时至今日,我方仍未收到你方的意见或批准。

因此,特通知你方,我方正在蒙受工期延误,其原因是与第25.4.6条款下的事件有关。依据合同第26条/补充条款S7[视情况取舍],我方有权,确实有权对此提出损失和(或)费用的补偿要求。合同条款要求的进一步的具体细节和计算依据将会尽快提交给你方。

你忠诚的

Dear Sir

We refer to our letter of the [*insert date*] enclosing two copies of drawings numbered [*insert numbers*] which we submitted for comments or approval. Such comments or approval should have been given within 5 [*or such number as is stated in the Employer's Requirements*] working days of the date of our letter. At the date of this letter, we have not received such comments or approval.

We, therefore, hereby notify you that we are suffering delay and the cause is a relevant event under clause 25.4.6 and a matter for which we are entitled to, and do, make application for reimbursement of loss and/or expense under clause 26/supplementary provision S7 [*delete as appropriate*] of the contract. Further particulars and estimates as required under the provisions of the contract will be submitted as soon as practicable.

Yours faithfully

函件 61

若发现文件之间有矛盾,致建筑师

本函不适用于 WCD98 合同

Letter 61

To architect, if discrepancy found between documents

This letter is not suitable for use with WCD 98

尊敬的先生:

  依据第 2.3 条 [当使用 IFC98 合同时,为"第 1.4 条";当使用 MW98 合同时,为 "4.1 条";当使用 GC/Works/1(1998) 合同时,为"第 2.3 条"],我方想提请你方注意已发现的下列矛盾之处[当使用 IFC98 合同或 MW98 合同时,代之以"不一致"]。

  [开列清单,给出工程量表参考项的准确细节,图号或建筑师指令的编号和日期]

  为避免我方施工进度的延误或中断,我方要求你方在[填入日期]前给出指令。

你忠诚的

Dear Sir

In accordance with clause 2.3 [*substitute* " 1.4" *when using IFC 98*, "4.1" *when using MW 98 or* "2(3)" *when using GC/Works/1 (1998)*] we bring to your attention the following discrepancies [*substitute* "inconsistencies" *when using IFC 98 or MW 98*] which we have discovered:

[*List, giving precise details of bills of quantities references, drawing numbers or dates and numbers of architect's instructions*]

In order to avoid delay or disruption to our progress, we require your instructions by [*insert date*].

Yours faithfully

函件 62
若发现"雇主要求"文件内有矛盾,致建筑师
本函仅适用于 WCD98 合同

Letter 62
To architect, if discrepancy within the Employer's Requirements
This letter is only suitable for use with WCD 98

尊敬的先生:

我方发现"雇主要求"文件内有矛盾[给出详细内容]。
我方"承包商方案"不涉及此事,因此,我方建议作如下修正:

[详细描述如何解决此项矛盾,包括提供图纸]

依据合同条件第 2.4.1 条,请书面批准我方的方案,或者提供另一个有详细内容的方案,上述二者的任何一个结果都将属于对"雇主要求"的变更。为了避免对工程的延误,我方要求在[填入日期]之前得到你方的批准或决定。
请把本函作为依据第 2.4.3 条发出的通知书。

你忠诚的

Dear Sir

We have found a discrepancy in the Employer's Requirements [*give details*]. Our Contractor's Proposals do not deal with the matter and, therefore, we propose the following amendment:

[*Describe in detail how to deal with the discrepancy including the provision of any drawings*]

In accordance with clause 2.4.1 of the conditions of contract, please either let us have your written agreement to our proposal or details of your alternative decision, either of which will rank as a change to the Employer's Requirements. We need your agreement or decision by [*insert date*] in order to avoid delay to the works.
Please take this as notice under clause 2.4.3.

Yours faithfully

函件 63

若"承包商方案"文件中有不一致之处,致建筑师
本函仅适用于 WCD98 合同

Letter 63

To architect, if discrepancy within the Contractor's Proposals
This letter is only suitable for use with WCD 98

尊敬的先生:

我方发现"承包商方案"文件中有不一致之处[给出详细内容]。我方提议用下列修正来消除这些矛盾之处。

[详细说明如何解决矛盾,包括提供图纸]

依据合同条件第 2.4.2 条,请决定你方愿意选择的不一致之处的那一项或接受我方建议的修正内容。为了避免工程的延误,我方要求在[填入日期]之前得到你方的书面决定或批准。

请把本函作为依据第 2.4.3 条发出的通知书。

你忠诚的

Dear Sir

We have found a discrepancy in our Contractor's Proposals [*give details*]. We propose the following amendment to remove the discrepancy:

[*Describe in detail how to deal with the discrepancy including the provision of any drawings*]

In accordance with clause 2.4.2 of the conditions of contract, please either decide which of the discrepant items you prefer or accept our proposed amendment. We require your decision or acceptance in writing by [*insert date*] in order to avoid delay to the works.
Please take this as notice under clause 2.4.3.

Yours faithfully

函件 64

若发现"雇主要求"和"承包商方案"文件之间有矛盾,致建筑师

本函仅适用于 WCD98 合同

Letter 64

To architect, if discrepancy found between Employer's Requirements and Contractor's Proposals

This letter is only suitable for use with WCD 98

尊敬的先生:

  我方发现在"雇主要求"和"承包商方案"之间有相互矛盾之处[给出详细内容]。合同中没有明确涉及如何处理这类情况。依据 JCT 合同的使用说明,这种情况不应发生。如果你方同意接受在"承包商方案"中处理这类事件的办法,就没有问题,不然我方提议召开一次现场会议。请尽快/按急件[视情况取舍]给予答复。

你忠诚的

Dear Sir

We have found a discrepancy between the Employer's Requirements and our Contractor's Proposals [*give details*]. The contract does not expressly deal with this situation. According to the JCT Practice Note, the situation should not occur. If you are content to accept the way we have dealt with the matter in our Contractor's Proposals, there is no problem, otherwise we suggest that a meeting on site is required. Please let us have your response as soon as possible/as a matter of urgency [*delete as appropriate*].

Yours faithfully

函件 65

致建筑师,说明法律规定和其他文件之间的不一致

本函不适用于 WCD98 合同

Letter 65

To architect, noting divergence between statutory requirements and other documents

This letter is not suitable for use with WCD 98

尊敬的先生:

  我方发现在法律规定和你方的图纸/详图/进度计划/工程量表/技术规范 [视情况取舍,并加上图号及工程量表中的页码和条目等等]之间有下列抵触之处。

  [填入相互抵触的细节]

  请在[填入日期]前给我们发出指令[注意,JCT98 合同允许建筑师在收到通知之日起的 7 天内发出指令]。若在上述日期之前不能得到你方指令,将会导致工程延误及中断,对此,我方将寻求恰当的经济补偿。

你忠诚的

Dear Sir

We have found what appears to be a divergence between statutory requirements and your drawing/detail/schedule/bills of quantities/specification [*delete as appropriate and add number of drawing, page and item number of bills of quantities, etc.*] as follows:

[*Insert details of the divergence*]

Please let us have your instruction by [*insert date-note that JCT 98 allows the architect 7 days from receipt of the notice to respond with an instruction*]. Failure to issue your instruction by that date will result in delay and disruption to the works for which we will seek appropriate financial recompense.

Yours faithfully

函件 66
致建筑师,说明法律规定与其他文件之间的矛盾
本函仅适用于 WCD98 合同

Letter 66
To architect, noting divergence between statutory requirements and other documents
This letter is only suitable for use with WCD 98

尊敬的先生:

我方发现法律规定和"雇主要求"/"承包商方案"[视情况取舍]之间有如下相互矛盾之处:

[填入相互矛盾的详细内容]

我方提议下列修正来解决这些矛盾:

[详细描述如何解决这些矛盾,包括提供图纸]

为了避免工程的延误,请在[填入日期]之前书面同意我方方案。请注意我方对合同文件的修改是依据合同条件第6.1.2条进行的,并请向我方提供一份拷贝。

你忠诚的

Dear Sir

We have found what appears to be a divergence between statutory requirements and the Employer's Requirements/our Contractor's Proposals [*delete as appropriate*] as follows:

[*Insert details of the divergence*]

We propose the following amendment to remove the divergence:

[*Describe in detail how to deal with the divergence including the provision of any drawing*]

Please let us have your written consent to our proposal by [*insert date*] in order to avoid delay to the works. Please note the amendment on the contract documents in accordance with clause 6.1.2 of the conditions of contract and send us a copy.

Yours faithfully

函件 67
当要求处理符合法律法规规定的突发事件时,致建筑师
本函不适用于 MW98 合同或 GC/工程/1(1998)合同

Letter 67
To architect, if emergency compliance with statutory requirements required
This letter is not suitable for use with MW 98 or GC/Works/1 (1998)

尊敬的先生:

　　依据合同条件第 6.1.4.2 条 [当使用 IFC98 合同时,为"第 5.4.2 条";当使用 WCD98 合同时,为"第 6.1.3.2 条"]的要求,我方在此提醒,我方有义务处理符合法律法规规定的突发事件。我方正在采取/已采取[视情况取舍]的紧急措施和步骤如下:

　　[填入细节]

　　我方认为已实施的工程和已提供的材料构成一个项变更事项[使用 CD81 合同时,代之以"变更"],并且我方将乐意收到任何进一步的必要指令。
　　依据第 25.2.1 条 [当使用 IFC98 合同时,为"第 2.3 条";当使用 WCD98 合同时,为"第 25.2.1 条"],请把本函作为工期延误的通知和依据第 26.1 条 [当使用 IFC98 合同时,为"第 4.11 条"]提出损失和(或)费用的补偿申请。

你忠诚的

Dear Sir

We hereby give notice as required by clause 6.1.4.2 [*substitute "5.4.2" when using IFC 98 or "6.1.3.2" when using WCD 98*] of the conditions of contract that we have been obliged to carry out work constituting emergency compliance with statutory requirements. The emergency and the steps we are taking/have taken [*delete as appropriate*] are as follows:

[*Insert details*]

We confirm that the work carried out and materials supplied rank as a variation [*substitute "change" when using CD 81*] and we should be pleased to receive any further instructions necessary.
Please take this letter as a notice of delay under clause 25.2.1.1 [*substitute "2.3" when using IFC 98 or "25.2.1" when using WCD 98*] and an application for loss and/or expense under clause 26.1 [*substitute "4.11" when using IFC 98*].
Yours faithfully

函件 68
若基准日后法律规定有变化时,致建筑师
本函仅适用于 WCD98 合同

Letter 68
To architect, if a change in statutory requirements after base date
This letter is only suitable for use with WCD 98

尊敬的先生:

在[填入日期]法律规定发生了变化,影响我方实施"承包商方案"[给出详细内容]。

我方正在做必要的修正,依据合同条件第 6.3.1 条,此修正应当作变更指令,一旦我方能把此变更估价出来,我会立即致函你方。依我方的意见,这不是一个依据补充条 S6 要处理的变更指令。因此,请把本函作为依据第 25.2.1 条发出的工期延误的通知和依据第 26.1 条发出的对损失和(或)费用的补偿申请。

你忠诚的

Dear Sir

On [*insert date*] there was a change in statutory requirements which affects our Contractor's Proposals [*give details*].

We are putting the necessary amendment in hand and we will write to you again as soon as we are in a position to value the amendment which is to be treated as a change instruction under the provisions of clause 6.3.1 of the conditions of contract. In our view, this is not an instruction to be dealt with under supplementary provision S6. Therefore, please take this letter as notice of delay under clause 25.2.1 and application for loss and/or expense under clause 26.1.

Yours faithfully

函件 69
若基准日后决定控制开发项目时,致建筑师
本函仅适用于 WCD98 合同

Letter 69
To architect, if development control decision after base date
This letter is only suitable for use with WCD 98

尊敬的先生:

　　我方刚收到[写明有关当局,如:当地规划局]的许可/批准[视情况取舍],并附上复印件一份。为了符合这些要求,有必要按下列内容[给出详细内容]修正我方的"承包商方案"。

　　我方正在做必要的修正,依据合同条件第 6.3.2 条,把此修正作为变更指令,一旦我方完成对此变更的估价,我方会立即致函你方。依我方的意见,这不是一个依据补充条 S6 要处理的变更指令。因此,请把本函作为依据第 25.2.1 条发出的工期延误的通知和依据第 26.1 条发出的对损失和(或)费用的补偿申请。

你忠诚的

Dear Sir

We have just received a permission/approval [*delete as appropriate*] from [*state the relevant authority, e.g. local planning authority*] and a copy is enclosed. To make them conform, it will be necessary to amend our Contractor's Proposals as follows: [*give details*]

We are putting the necessary amendment in hand and we will write to you again as soon as we are in a position to value the amendment which is to be treated as a change instruction under the provisions of clause 6.3.2 of the conditions of contract. In our view, this is not an instruction to be dealt with under supplementary provision S6. Therefore, please take this letter as notice of delay under clause 25.2.1 and application for loss and/or expense under clause 26.1.

Yours faithfully

函件 70
致雇主,反对任命建筑师的替代人选
本函仅适用于 JCT98 合同或 IFC98 合同

Letter 70
To employer, objecting to the nomination of a replacement architect
This letter is only suitable for use with JCT 98 or IFC 98

尊敬的先生:

　　依据第3条的规定,在此正式通知,我方反对任命由[填入地址]的[填入姓名]更换[填入前任建筑师的姓名和地址],成为本合同的建筑师。
　　我方反对的理由是[填入反对的具体理由]
　　建筑师与承包商之间的良好工作关系对任何项目的顺利竣工都是至关重要的。基于此点,我方期待你方能重新考虑此任命。

你忠诚的

Dear Sir

Under the provisions of article 3 we hereby formally give notice of our objection to the nomination of [*insert name*] of [*insert address*] as architect for the purpose of this contract in succession to [*insert name and address of previous architect*].

The grounds for our objections are [*insert particular reasons for objection*].

A good working relationship between architect and contractor is vital to the successful completion of any project. With this in mind, we look forward to hearing that you have reconsidered the nomination.

Yours faithfully

函件 71
若建筑师的替代人选尚未任命时,致雇主
本函不适用于 WCD98 合同或 GC/Works/1(1998)合同

Letter 71
To employer, if replacement architect not appointed
This letter is not suitable for use with WCD 98 or GC/Works/1 (1998)

尊敬的先生:

我方接到通知/被告知[视情况取舍]合同中命名的建筑师已于或约在[填入日期]终止履行他的职责。依据第3条的规定,你方有责任在21天之内[当使用IFC98合同或MW98合同时,为"14天"]任命一个建筑师的替代人选。

现在,在你方应作出任命的时限之后已过约[填入数量]天,但你方仍未作出任命。其结果就是,最基本的工作如指令或证书没人签发,工期延期条款不能得到实施。因此,当我方需要得到继续施工的指令时,我方却不得不停下那项具体工作,为了做好记录,我方只能通知你方要求得到继续施工的指令。我方面对的现实是,整个工地将很快停工,我方必须考虑自身的处境。无论如何,工期很快将成为最大的问题。

任何替代的建筑师都需要时间来熟悉了解本项目的情况,我方强烈要求你方立即作出任命。同时,由于你方的违约,你方应承担由此造成我方蒙受损害、损失和费用的责任。

你忠诚的

Dear Sir

We were notified/became aware[*delete as appropriate*] that the architect named in the contract had ceased to act on or about the [*insert date*]. Under the provisions of article 3 you are obliged to nominate a replacement architect within 21 [*substitute "14" when using IFC 98 or MW 98*] days.

It is now some[*insert number*] days since you should have made the appointment and you have failed to do so. The result of this is that, at the most basic level, no instructions or certificates can be issued and the extension of time provisions cannot be operated. Therefore, as soon as we require an instruction to enable us to proceed, we shall be obliged to stop that particular activity although we will, for the record, notify you of the instruction required. We envisage that the whole site will be at a standstill very shortly and we must consider our position. In any event, time will soon become at large.

Any replacement architect will need time to absorb the detail of this project and we urge you to make the appointment immediately. In the meantime you are liable to us for all the damage, loss and expense caused to us by your breach.

Yours faithfully

函件 72

致雇主,反对工料测量师替代人选的任命
本函仅适用于 JCT98 合同或 IFC98 合同

Letter 72

To employer, objecting to the nomination of a replacement quantity surveyor
This letter is only suitable for use with JCT 98 or IFC 98

尊敬的先生:

　　依据第 4 条规定,我方在此正式通知你方,我方反对任命位于[填入地址]的[填入姓名]来替换[填入原先工料测量师的地址和姓名]作为本合同的工料测量师。
　　我方反对的理由是[填入反对的具体理由]。
　　工料测量师与承包商之间的良好工作关系对顺利完成项目是至关重要的。基于此点,我方期待你方重新考虑该任命。

你忠诚的

Dear Sir

Under the provisions of article 4 we hereby formally give notice of our objection to the nomination of [*insert name*] of [*insert address*] as quantity surveyor for the purpose of this contract in succession to [*insert name and address of previous quantity surveyor*].

The grounds for our objection are [*insert particular reasons for objection*].

A good working relationship between quantity surveyor and contractor is vital to the successful completion of any project. With this in mind, we look forward to hearing that you have reconsidered the nomination.

Yours faithfully

函件 73

致建筑师,关于工程管理人员在现场签发的指示

本函不适用于 GC/Works/1(1998)合同

Letter 73

To architect, regarding directions issued on site by the clerk of works

This letter is not suitable for use with GC/Works/1 (1998)

尊敬的先生:

　　工程管理人员已于[填入日期]在现场签发了[填入编号]指示,现附上复印件一份。

　　当然,这些指示没有合同效力[当使用 JCT98 合同时,代之以"除非在该指示签发后的二个工作日内得到你方的确认"]。很明显,工程管理人员就缺陷工程的整改指示是很有帮助的。我方十分希望在现场避免误解,基于这种精神,我方建议,除了这类有关缺陷工程整改指示之外,工程管理人员不要签发任何其他指示。他可以用电话直接向你汇报其他的事情合适的建筑师的指令的签发由你方决定。

　　依据我方的看法,上述做法必将消除很多由目前做法必然会带来的不确定因素。期待得到你的意见。

你忠诚的

Dear Sir

The clerk of works has issued direction number [*insert number*] dated [*insert date*] on site, a copy of which is enclosed.

Such directions have, of course, no contractual effect [*insert* "*unless you confirm them within 2 working days of issue*" *when using JCT 98*]. Clearly, the directions of the clerk of works issued in relation to the correction of defective work can be very helpful. We are anxious to avoid misunderstandings on site and in this spirit we suggest that the clerk of works should issue no further directions, other than those relating to defective work. He can refer all other matters directly to you by telephone and, at your discretion, a proper architect's instruction can be issued.

In our view, the above system would remove a good deal of the uncertainty which must result from the present state of affairs. We look forward to hearing your comments.

Yours faithfully

函件 74

致建筑师,关于工程管理人员在现场签发的指令
本函仅适用于 GC/Works/1(1998)合同

Letter 74

To architect, regarding instructions issued on site by the clerk of works
This letter is only suitable for use with GC/Works/1 (1998)

尊敬的先生:

  工程管理人员于[填入日期]在现场签发了[填入编号]指令,现附上复印件一份。

  合同条件第4(2)条规定,除非依据第4(4)条的规定,你方明确书面给工程管理人员其他授权并且通知我方,否则他只能签发依据第31条规定范围内的指令。

  附上的指令不属于第31条规定的范围,并且我方尚未收到你方依据第4(4)条签发的通知。我方现场的代理人已接到我方的严格指示,除非合同明确授权,否则,工程管理人员的指令将不会得到执行。若你方认为该指令应执行,请立即给我方一个确认,或者,给我方一个依据第4(4)条的通知:工程管理人员享有合同授予你的同等权力来签发任何指令。

你忠诚的

Dear Sir

The clerk of works has issued instruction number [*insert number*] dated [*insert date*] on site, a copy of which is enclosed.

Clause 4(2) of the conditions of contract provides that the clerk of works may only issue instructions under clause 31 unless you expressly delegate other powers to him in writing and notify us under the provisions of clause 4(4).

The enclosed instruction does not fall under the provisions of clause 31 and we have not yet received a clause 4(4) notice from you. Our agent on site has strict instructions from us that instructions of the clerk of works are not to be executed unless expressly empowered by the contract. If you wish the instruction to be carried out, please let us have your confirmation without delay or, alternatively, a notice under clause 4(4) that the clerk of works has power to issue any instructions which the contract empowers you to issue.

Yours faithfully

函件 75
致建筑师,关于工程管理人员毁坏工程或材料

Letter 75
To architect, if clerk of works defaces work or materials

尊敬的先生:

工程管理人员把他认为有缺陷的工程和材料毁坏是常发生的举动。这种行为的目的为让缺陷引起承包商注意,并且确保不能对该缺陷熟视无睹。

基于下列原因,我方反对这种做法。

1)如此标示的工程或材料可能并无缺陷,而我方却要做额外修补工作,而雇主在这种情况下要承受额外的成本支出。

2)这样标示的工程或材料,若真正是有问题的,将得不到付款,当被拆除时,将属于我方的财产。我方可以把它用在执行不同标准的其他项目之中。工程管理人员的毁坏将造成它们不能被再利用。

对我方今天在现场发现的这些毁坏标记,我方将不予追究,但若这种事情继续发生,对每次事件我方将寻求经济补偿。

你忠诚的

Dear Sir

It is common practice for the clerk of works to deface work or materials which he considers to be defective. The basis for such action is to bring the defect to the notice of the contractor and ensure that it cannot remain without attention.

We object to the practice on the following grounds:

1. The work or materials so marked may not be defective and we will be involved in extra work and the employer in extra costs in such circumstances.

2. The work or materials so marked, if indeed defective, will not be paid for and will be our property when removed. We may be able to incorporate it in other projects where a different standard is required. Defacement by the clerk of works would prevent such re-use.

We will take no point about the defacing marks we noted on site today, but if the practice continues, we will seek financial reimbursement on every occasion.

Yours faithfully

函件 76
当若干"专家"工作人员要参观工地时,致建筑师
本函不适用于 WCD98 合同或 MW98 合同

Letter 76
To architect, if numerous "specialist" clerks of works visiting site
This letter is not suitable for use with WCD 98 or MW 98

尊敬的先生:

  合同条件第 12 条 [当使用 IFC98 合同时,为"第 3.10 条";当使用 GC/Works/1 (1998)合同时,为"第 4(2)条"]允许雇主任命工程管理人员。你方[填入日期]的来函通知,[填入姓名]将是工程管理人员。在过去的二周中,许多人员来我方现场办公室自称是"专家工作人员"[或用实际的头衔替代]并要求进入现场。

  这些人员却既不为我方所知,合同中也没有提及他们,因此,我方已履行我方的权力,不让他们进入工地。

  我方已经允许[填入日期]的项目开工会上介绍的那些咨询工程师检查工地,虽然合同上并没有提到此事,而我方想表示我方全面的合情合理的合作。我方认为,允许"专家工作人员"自由进入工地是不合理的。如果你方要求我方允许他们享有这样的待遇,请书面通知我方,但请注意,在这种情况下,我方认为我方的进度可能会受到影响,我方将就我方权利和补偿采取合适的法律措施。

                       你忠诚的

Dear Sir

Clause 12 [*substitute* "3.10" *when using IFC 98 or* "4(2)" *when using GC/Works/1 (1998)*] of the conditions of contract permits the employer to appoint a clerk of works. Your letter of the [*insert date*] informed us that the clerk of works would be [*insert name*]. During the past two weeks a number of persons have presented themselves at our site office purporting to be "specialist clerks of works" [*or substitute the actual title*] and demanding access to the works.

These persons are neither known to us nor included in the contract and, therefore, we have exercised our right to exclude them from the works.

We already permit inspections by those consultants introduced at the project start meeting of the [*insert date*] even though the contract is silent as to their existence, as we desire to extend all reasonable co-operation. To allow "specialist clerks of works" free access would not, in our view, be reasonable. If you require us to grant them such access, please so inform us in writing, but note that, in such circumstances, we consider that our progress would be hindered and we will take appropriate legal advice in regard to our rights and remedies.

Yours faithfully

函件 77
当工程管理人员直接指挥施工人员时,致建筑师

Letter 77
To architect, if clerk of works instructs operatives direct

尊敬的先生:

能允许我方提请你注意在工地上发生着一种令人遗憾的情况吗?我方与工程管理人员一直保持良好的合作关系,但他们现正养成一种习惯,在工地上直接对我方的施工人员发布口头指示,甚至训斥他们中的一些人技术差。

若把这些问题告知负责人/现场经理[视情况取舍]我方将非常重视工程管理人员的这些意见。正如可预见到的,目前的情况将导致一定程度的混乱,使我方不得不浪费很多宝贵的时间来梳理这些弄乱了的事情。我方已非正式地向工程管理人员提出过,但无效果。我方渴望避免在现场发生紧张关系,这对谁都无益,若你方能尽快处理此事,我方将不胜感激。

你忠诚的

Dear Sir

May we draw your attention to an unfortunate situation which is developing on site? The clerk of works, with whom we have had the most cordial relations, is getting into the habit of giving oral directions to our operatives on site even going so far as to reprimand some of them for poor workmanship.

We very much value all the comments of the clerk of works, provided that they are addressed to the person-in-charge/site manager [*aelete as appropriate*]. The present situation is causing, as one might expect, a degree of disruption as we have to waste valuable time smoothing ruffled feathers. We have spoken to the clerk of works informally, but to no effect. We are anxious to avoid tension on the site which would be in no one's best interests and we should be grateful if you would deal with this matter as soon as possible.

Yours faithfully

函件 78

致工料测量师,提交价格报表
本函不适用于 MW98 合同或 GC/Works/1(1998)合同

Letter 78

To quantity surveyor, submitting a price statement
This letter is not suitable for use with MW 98 or GC/ Works/ 1 (1998)

尊敬的先生:

鉴于我方于[填入日期]收到的[填入日期]签发的[填入编号]指令,我方依据第13.4.1.2条[当使用 IFC98 合同时,为"第3.7.1.2条",当使用 WCD98 合同时,为"第12.4.2条"]附上我方的价格报表。

报表表明了依据第13.5条[当使用 IFC98 合同时,为"第3.7.2条""第3.7.10条",当使用 WCD98 合同时,为"第12.5条"]基础上计算出来的我方的工程价格。

[若合适,加上:]

我方另外附上了补偿损失和(或)费用的一笔总额要求以及我方对该工程竣工时间的调整。

[然后:]

期待你方在收到本函的 21 天内给予答复。

你忠诚的

Dear Sir

Further to receipt of instruction number [*insert number*] dated [*insert date*] which we received on the [*insert date*], we enclose our price statement for compliance under the provisions of clause 13.4.1.2 [*substitute* "*3.7.1.2*" *when using IFC 98 or* "*12.4.2*" *when using WCD 98*].

It states our price for the work, which has been calculated on the basis of clause 13.5 [*substitute* "*clauses 3.7.2 to 3.7.10*" *when using IFC 98 or* "*clause 12.5*" *when using WCD 98*].

[*If appropriate, add*:]

We have separately attached our requirements for an amount to be included in lieu of ascertainment of loss and/or expense and our adjustment to the time for the completion of the Works.

[*Then*:]

We look forward to a response within 21 days of your receipt of this letter.

Yours faithfully

函件 79

当价格报表没有全部被接受时,致工料测量师

本函不适用于 MW98 合同或 GC/Works/1(1998)合同

Letter 79

To quantity surveyor if price statement not accepted in full

This letter is not suitable for use with MW 98 or GC/Works/1 (1998)

尊敬的先生:

你方 [填入日期] 来函今日收到,谢谢。来函告知我方的部分价格没有被你方接受。

[选择加入下述两段中的一段]

在你方的通知中没有详述不接受的理由,这是不符合第 13.4.1.2A4.1 条款[当使用 IFC98 合同时,为"第 3.7.1.2A4.1 条";或当使用 WCD98 合同时,为"第 12.4.2A4.1 条"]规定的。

[或]

你方不提供经修正的价格报表,这是不符合第 13.4.1.2A4.1 条[当使用 IFC98 合同时,为"第 3.7.1.2A4.1 条";或当使用 WCD98 合同时,为"第 12.4.2A4.1 条"]规定的。

[然后加上]

只有当你方提供了经修正的价格报表,我方是否接受部分或全部的修正价格报表的 14 天期限才可以起算。

你忠诚的

Dear Sir

Thank you for your letter dated [*insert date*] received today, in which you notified us that part of our price statement is not accepted.

[*Add either*:]

Contrary to clause 13.4.1.2A4.1 [*substitute* "3.7.1.2A4.1" *when using IFC 98 or* "12.4.2A4.1" *when using WCD 98*], you failed to include in your notification your detailed reasons for not accepting.

[*Or*:]

Contrary to clause 13.4.1.2A4.1 [*substitute* "3.7.1.2A4.1" *when using IFC 98 or* "12.4.2A4.1" *when using WCD 98*], you failed to supply an amended price statement.

[*Then add*:]

Until you do so, the 14 days in which we must state whether we accept any or all of the amended price statement does not begin to run.

Yours faithfully

函件 80

致工料测量师,提交按第 13A 条的报价
本函仅适用于 JCT98 合同

Letter 80

To quantity surveyor, submitting a clause 13A quotation
This letter is only suitable for use with JCT 98

尊敬的先生:

  依据第 13A.1.2 条的规定以及遵照[填入日期]签发的[填入编号]指令的要求,在此附上我方的报价[若合适,加入:"并包括了下列指定分包商的报价",然后列出有关的指定分包商]。报价包含了要求的合同总价调整,并附计算书;要求的工程竣工日期的调整;依据第 26 条所作调整后发生的损失和/或费用的金额;以及编制报价的费用。

[若指令有特殊要求时,应加入:]

说明性资料将包括所需的额外资源及实施工程的方法。

[然后加上]

本报价从收到日起的 7 天内有效。

你忠诚的

Dear Sir

Under the provisions of clause 13A.1.2 and in compliance with instruction number [*insert number*] dated [*insert date*], we enclose our quotation [*if appropriate add*: "*which includes quotations from the following nominated sub-contractors:*" *then list the relevant nominated sub-contractors*]. The quotation comprises the required adjustment to the contract sum with calculations, the required adjustment to the time for completion of the Works, the amount of loss and/or expense required in lieu of any adjustment under clause 26 and the cost of preparing the quotation.

[*If the instruction specifically requires it add*:]

Indicative information is included about the extra resources needed and the method of carrying out the work.

[*Then add*:]

The quotation is open for acceptance for 7 days from the date of receipt.

Yours faithfully

函件 81

致建筑师,有关计日工作凭证的核实
本函仅适用 JCT98 合同

Letter 81

To architect, regarding verification of vouchers for daywork
This letter is only suitable for use with JCT 98

尊敬的先生:

  我方在此附上有关[填入工程内容]的[填入编号]收款凭证。若你方能依据合同条件第 13.5.4 条的要求进行核实,我方将不胜感激。

  [若合适,加入:]

  工程管理人员总是要求查看这些凭证。若你能通知我方,该工程管理人员是你方的授权代表来贯彻第 13.5.4 条要求的,那么对我方而言可以节省时间,他就能依据合同正式确认这些凭证。我方有兴趣知道这个建议能否得到你方的批准。

你忠诚的

Dear Sir

We enclose vouchers numbers [*insert numbers*] in respect of [*insert description of work*]. We should be pleased if you would verify them as required by clause 13.5.4 of the conditions of contract.

[*If appropriate, add*:]

The clerk of works always asks to see such vouchers and it occurred to us that it would save time if you notified us that the clerk of works was your authorised representative for the purpose of clause 13.5.4. He could officially verify the vouchers as required by the contract. We should be interested to hear whether this suggestion meets with your approval.

Yours faithfully

函件 82
若对工程是否属于变更或包括在合同内产生不同意见时，致雇主
专递/挂号邮件
毫无偏见
Letter 82
To employer, if disagreement over whether work is a variation or included in the contract
Special / recorded delivery
WITHOUT PREJUDICE

尊敬的先生：

　　基于你方[填入日期]的来函及我方[填入日期]有关[描述工程内容]的复函，我方确信这项工程不包括在合同之内，因此，该工程属变更，对此，我方有权得到付款。

　　我方知道，在你方正式授权签发变更指令之前，我方可以拒绝进一步实施该工程。若你方继续拒绝作出这样的授权，我方可把已签合同作为一个无法履行的合同，并进行诉讼要求赔偿损失。

　　在无损我方权利的前提下，暂把争议搁置起来，等将来由裁决或仲裁解决此事，或者如果你方书面同意并确认你方将不否认我方有权要求支付；在合同的实际施工中，确认工程不在合同范围之内，那么，我方就准备实施此工程。

你忠诚的

抄送：建筑师

Dear Sir

We refer to your letters of the [*insert dates*] and ours of the [*insert dates*] relating to [*describe work*]. It is our firm view that this work is not included in the contract and, therefore, constitutes a variation for which we are entitled to payment.

We are advised that we can refuse further performance until you authorise a variation. If you continue to refuse to so authorise, we may treat the contract as repudiated and sue for damages.

Without prejudice to our rights, we are prepared to carry out the work, leaving this matter in abeyance for future determination by adjudication or arbitration, if you will agree in writing and confirm that you will not deny our entitlement to payment in such reference if, on the true construction of the contract, the work is held to be not included.

Yours faithfully
Copy: Architect

函件 83
致建筑师,要求说明授权签发指令的条款
本函不适用 MW98 合同或 GC/Works/1(1998)合同

Letter 83
To architect, requiring him to specify the clause empowering an instruction
This letter is not suitable for use with MW 98 or GC/ Works/ 1 (1998)

尊敬的先生:

今天我方收到了你方于[填入日期]签发的[填入编号]指令,要求我方[填入指令内容]。

依据合同条件第4.2条[当使用IFC98合同时,为"第3.5.2条"]我方要求你方书面说明授权签发上述指令的合同规定。

你忠诚的

Dear Sir

We have received today your instruction number [*insert number*], dated [*insert date*] requiring us to [*insert substance of instruction*].

We request you, in accordance with clause 4.2 [*substitute* " *3. 5. 2* " *when using IFC 98*] of the conditions of contract, to specify in writing the provision which empowers the issue of the above instruction.

Yours faithfully

函件 84 致建筑师,确认口头指令 本函仅适用 JCT98 合同或 WCD98 合同 **Letter 84** To architect, confirming an oral instruction *This letter is only suitable for use with JCT 98 or WCD 98*

尊敬的先生:

　　我方在此确认,你方于[填入日期]口头指令我方实施[填入指令详细内容]。依据合同条件第4.3.2条规定,你有7天时间可用书面否认此事。

　　[或]

　　你方于[填入日期]口头指示我方实施[填入指令的详细内容]。我方执行了你方的指令,正如你方所知,因为[填入原因],你方并没有依据合同条件第4.3.2条书面确认你方的指令。因此,若你方能依据第4.3.2.2条,行使你方的权力并书面确认你方的指令,我方将不胜感激。

你忠诚的

Dear Sir

We hereby confirm that on [*insert date*], you orally instructed us to [*insert the instruction in detail*]. Under the provisions of clause 4.3.2 of the conditions of contract, you have 7 days in which to dissent in writing.

[*Or:*]

On [*insert date*], you orally instructed us to [*insert the instruction in detail*]. We complied with your instruction, as you are aware, but neglected to confirm your instruction as provided by clause 4.3.2 of the conditions of contract because [*insert reason*]. We should be pleased, therefore, if you would exercise your power under clause 4.3.2.2 and confirm the instruction in writing.

Yours faithfully

函件 85

致建筑师，要求确认口头指令
本函仅适用于 GC/Works/1(1998)合同

Letter 85

To architect, requesting confirmation of an oral instruction
This letter is only suitable for use with GC/Works/1 (1998)

尊敬的先生：

　　你方[填入日期]口头指令我方实施[填入指令的详细内容]。依据合同条件第40(3)条，若能收到你方的书面确认，我方将不胜感激。

你忠诚的

Dear Sir

On [*insert date*], you orally instructed us to [*insert the instruction in detail*]. In accordance with clause 40(3) of the conditions of contract, we should be pleased to receive your written confirmation.

Yours faithfully

函件 86
若口头指令未得到书面确认时,致建筑师
本函仅适用 MW98 合同

Letter 86
To architect, if oral instruction not confirmed in writing
This letter is only suitable for use with MW 98

尊敬的先生:

  你方[填入日期]口头指令我方实施[填入指令的详细内容]。我方已执行了你方的指令,正如你所知,我方尚未收到依据合同条件第 3.5 条规定你方应签发的书面确认。若你方能立即提交我方书面确认,我方将不胜感激。

你忠诚的

Dear Sir

On [*insert date*], you orally instructed us to [*insert the instruction in detail*]. We have carried out your instruction, as you are aware, but we have not yet received written confirmation as provided by clause 3.5 of the conditions of contract. We should be pleased, therefore, if you would send us your written confirmation immediately.

Yours faithfully

函件 87 致建筑师,反对改变责任或规定的指令 本函不适用 MW98 合同或 GC/Works/1(1998)合同 **Letter 87** To architect, objecting to instruction varying obligations or restrictions *This letter is not suitable for use with MW 98 or GC/Works/1 (1998)*

尊敬的先生:

你方[填入日期]签发的[填入编号]指令,今日收悉,谢谢。你方的指令要求变更[当使用 WCD98 合同时,代之以"更改"]第 13.1.2 条的内容[当使用 WCD98 合同时,为"第 12.1 条";或当使用 IFC98 合同时,为"第 3.6.2 条"]。

我方有理由不遵从你方的指令,因为[填入反对的理由]。

我方的反对是依据第 4.1.1.1 条[当使用 IFC98 合同时,为"第 3.5.1 条"]的规定,若能依据我方意见,撤销或修改你方指令,我方将不胜感激。

[若合适,加入:]

为了避免由于[依实际情况填入]造成工程的延误或中断,请在[填入日期]前答复我方。

你忠诚的

Dear Sir

Thank you for your instruction number [*insert number*], dated [*insert date*] which we received today. Your instruction requires a variation [*substitute "change" when using WCD 98*] within the meaning of clause 13.1.2 [*substitute "12.1" when using WCD 98 or "3.6.2" when using IFC 98*].

We have reasonable objection to complying with your instruction, because [*insert grounds of objection*].

Our objection is submitted under the provisions of clause 4.1.1.1 [*substitute "3.5.1" when using IFC 98*] and we should be pleased if you would withdraw or revise your instructions in the light of our comments.

[*Add, if appropriate*:]

Please let us have your reply by [*insert date*] in order to avoid the possibility of delay or disruption due to [*insert as appropriate*].

Yours faithfully

函件 88a
致建筑师,收到要求执行指令的 7 天通知
本函不适用于 GC/Works/1(1998)合同

Letter 88a
To architect, on receipt of 7 day notice requiring compliance with instruction
This letter is not suitable for use with GC/Works/1 (1998)

尊敬的先生:

  今天我方收到了你方[填入日期]的通知,这是你方依据合同条件第4.1.2条[当使用IFC98合同时,为"第3.5.1条";当使用MW98合同时,为"第3.5条"]认为应签发的指令。

  [任取下述三段之一加入]

  我方将立即执行你方[填入日期]编号为[填入编号]的指令。但是,这样的照办将不影响我方应保留的其他权利和我方应享有的赔偿。

  [或]

  我方认为我方已执行了你方[填入日期]编号为[填入编号]的指令。雇主任何要想雇用其他人员和(或)从我方的应付款项中或将付款项中扣除费用将是严重的违约行为,对此,我方将寻求合理的赔偿。在无损我方上述权利的前提下,若你方能立即撤除要求遵照执行的通知,[填入姓名]先生将高兴地在工地与您会面,以便解决这看来是令人遗憾的误解。

  [或]

  在你方规定的时间内按你方的要求来执行该指令,这是不合适且不现实的,因为[填入原因]。你方可放心,我方没有忘记在这件事上我方的责任和义务,一旦[描述工作内容]完成了,我方将立即执行你方[填入日期]编号为[填入编号]的指令。依据上述解释,若你方能立即撤除要求执行的通知,我方将十分高兴。若在[填入日期]之前我方没有收到你方的答复,我方将立即执行该指令,但请把本函作为我方的通知,如此紧急执行该项指令将是随后会发生的工期索赔以及损失和(或)费用索赔的理由。

你忠诚的

抄送:雇主

Dear Sir

We have today received your notice dated [*insert date*] which you purport to issue under the provisions of clause 4.1.2 [*substitute "3.5.1" when using IFC 98 or "3.5" when using MW 98*] of the conditions of contract.

[*Add either*:]

We will comply with your instruction number [*insert number*], dated [*insert date*] forthwith, but such compliance is without prejudice to and reserving any other rights and remedies which we may possess.

[*Or*:]

We consider that we have already complied with your instruction number [*insert number*], dated [*insert date*]. Any attempt by the employer to employ other persons and/or deduct from any monies due or to become due to us will be deemed to be a serious breach of contract for which we will seek appropriate remedies. Without prejudice to the foregoing, if you will immediately withdraw your notice requiring compliance, Mr [*insert name*] will be happy to meet you on site to sort out what appears to be an unfortunate misunderstanding.

[*Or*:]

It is not reasonably practicable to comply as you require within the period you specify, because [*insert reasons*]. You may be assured that we have not forgotten our obligations in this matter and we intend to carry out your instruction number [*insert number*], dated [*insert date*] as soon as [*indicate operation*] is complete. In the light of this explanation, we should be pleased to hear, by return, that you withdraw your notice requiring compliance. If we do not have your reply by [*insert date*], we will immediately comply, but take this as notice that such immediate compliance will give grounds for substantial claims for extension of time and loss and/or expense.

Yours faithfully

Copy: Employer

函件 88b
致建筑师,收到要求执行指令的通知
本函仅适用于 GC/Works/1(1998)合同
Letter 88b
To architect, on receipt of notice requiring compliance with instruction
This letter is only suitable for use with GC/Works/1 (1998)

尊敬的先生:

今天我方收到了你方[填入日期]的通知,你方认为该通知是依据合同条件第53条的规定而签发的。

[下述三段任选一段加入]

我方将执行你方[填入日期]签发的编号为[填入编号]指令。但这样的遵从将不影响我方保留我方应享有的其他权利和赔偿。

[或]

我方认为,我方已执行了你方的[填入日期]编号为[填入号码]指令。雇主想提供劳动力和(或)其他任何东西,或与其他人签订合同和(或)想从我方处得到所谓的额外好处或费用的企图都是严重的违约行为,对此,我方将寻求适当的赔偿。在上述权利不受到损害的前提下,若你方能立即撤销要求执行的通知,[填入姓名]先生将高兴地在工地现场与您会面,以便解决这看来是令人遗憾的误解。

[或]

在你方规定的时间内按你方的要求来执行该指令,这是不合适且不现实的,因为[填入原因]。你方可放心,我方没有忘记在这件事上我方的责任和义务,一旦[描述工作内容]完成了,我方将立即执行你方[填入日期]编号为[填入编号]的指令。依据上述解释,若你方能立即撤除要求执行的通知,我方将十分高兴。若在[填入日期]之前我方没有收到你方的答复,我方将立即执行该指令,但请把本函作为我方的通知,如此紧急执行该项指令将是随后会发生的工期索赔以及损失和(或)费用索赔的理由。

你忠诚的

抄送:雇主

Dear Sir

We have today received your notice dated [*insert date*] which you purport to issue under the provisions of clause 53 of the conditions of contract.

[*Add either*:]

We will comply with your instruction number [*insert number*], dated [*insert date*] forthwith, but such compliance is without prejudice to and reserving any other rights and remedies which we may possess.

[*Or*:]

We consider that we have already complied with your instructed number [*insert number*], dated [*insert date*]. Any attempt by the employer to provide labour and/or any things, or enter into a contract with others and/or recover alleged additional costs and expenses from us will be deemed to be a serious breach of contract for which we will seek appropriate remedies. Without prejudice to the foregoing, if you will immediately withdraw your notice requiring compliance, Mr [*insert name*] will be happy to meet you on site to sort out what appears to be an unfortunate misunderstanding.

[*Or*:]

It is not reasonably practicable to comply as you require within the period you specify, because [*insert reasons*]. You may be assured that we have not forgotten our obligations in this matter and we intend to carry out your instruction number [*insert number*], dated [*insert date*] as soon as [*indicate operation*] is complete. In the light of this explanation, we should be pleased to hear, by return, that you withdraw your notice requiring compliance. If we do not have your reply by [*insert date*], we will indeed immediately comply, but take this as notice that such immediate compliance will give grounds for substantial claims for extension of time and loss and/or expense.

Yours faithfully

Copy: Employer

函件 89

致建筑师，运走未使用的材料

本函不适用于 MW98 合同或 GC/Works/1(1998)合同

Letter 89

To architect, removal of unfixed materials

This letter is not suitable for use with MW 98 or GC/Works/1 (1998)

尊敬的先生：

[阐述货物或材料的数量及其名称]目前正存储在现场。我方的观点是这些材料应储存在[指出地点名称]，这是因为[说明原因]。若能收到你方依据合同条件第16.1条[当使用 WCD98 合同时，为"第15条"；当使用 IFC98 合同时，为"第1.10条"]的规定，书面同意把这些材料撤出现场，我方将十分高兴。

你忠诚的

Dear Sir

[*State quantity and nature of goods or materials*] are presently stored on site. It is our view that these materials should be stored at [*name place*] because [*state reason*]. We should be pleased to receive your written consent to the removal from site of these materials in accordance with the provisions of clause 16.1 [*substitute " 15" when using WCD 98 or " 1.10" when using IFC 98*] of the conditions of contract.

Yours faithfully

函件 90
若材料采购不到时,致建筑师
本函仅适用于 JCT98 合同及 WCD98 合同

Letter 90
To architect, if materials are not procurable
This letter is only suitable for use with JCT 98 and WCD 98

尊敬的先生:

  我方供应商[填入名称]告知,[填入材料名称]采购不到,原因是[填入理由]。现附上他们[填入日期]来函的复印件,供参阅。我方能采购到[填入材料名称]作为替代材料并能满足你方要求,它的价格为[填入价格]。我方希望在[填入日期]之前得到你方的指令,否则将有延误工程进度的风险。

你忠诚的

Dear Sir

We have been informed by our suppliers [*insert name*] that [*insert description of material*] is not procurable because [*insert reason*]. A copy of their letter dated [*insert date*] is enclosed for your information. We can obtain [*insert description of material*] which you may consider to be an alternative which meets your requirements at a cost of [*insert cost*]. We should be pleased to have your instructions by [*insert date*] because there is a danger that there will be delay to the progress of the works.

Yours faithfully

函件 91

当工程或材料或货物验收不合格时,致建筑师
本函仅适用于 IFC98 合同

Letter 91

To architect, after failure of work or materials or goods
This letter is only suitable for use with IFC 98

尊敬的先生:

[若建筑师已通知不合格,信函开头应为:]

你方[填入日期]通知我方[简单描述细节]的来函已收悉,谢谢。

[其他情况下,信函开头应为:]

我方不得不通知你方,在[填入日期,必须不超出本函日期前 7 天]我方发现了不符合合同要求的质量事件[简单描述质量不合格事件内容]。

[然后加入:]

依据第 3.13.1 条,我方建议采取[填入要采取确保不再发生同类事件的行动],确保在这些范围/材料/货物[视情况取舍]中不再发生同类错误。我方将十分高兴能立即收到你方对我方建议的书面批准。

你忠诚的

Dear Sir

[*If the architect has notified failure, begin*:]

Thank you for your letter of the [*insert date*], notifying us [*briefly give details*].

[*Otherwise, begin*:]

We have to notify you that a failure of [*briefly give details*] to be in accordance with the contract was discovered on [*insert date which must be not more than 7 days before the date of this letter*].

[*Then add*:]

In accordance with clause 3.13.1, we propose to [*insert action to be taken to ensure there are no similar failures*] to ensure that there is no similar failure in these areas/materials/goods [*delete as appropriate*]. We should be pleased to receive your immediate written approval of our proposals.

Yours faithfully

函件 92

若承包商反对执行按第 3.13.1 条发出的指令时,致建筑师

本函仅适用于 IFC98 合同

专递/挂号邮件

Letter 92

To architect, if contractor objects to complying with a clause 3.13.1 instruction

This letter is only suitable for use with IFC 98

Special/recorded delivery

尊敬的先生:

　　收到你方[填入日期]编号为[填入号码]指令,指示我方实施[填入工作的内容],这是你方依据合同条件第 3.13.1 条签发的指令。我方认为该指令是不合理的,因为[说明原因]。

　　若你方在收到本函后 7 天之内不用书面撤销该指令或修改该指令以便消除我方的反对意见,那么,对你方指令中要求的打开/测试[视情况取舍]的性质及范围在所有情况下是否合理都将存在争议和分歧。这样的争议或分歧将提交裁决/仲裁[视情况取舍]解决。在这种情况中,我方仍会履行我方的义务直至裁决/仲裁[视情况取舍]的结果和额外成本和工期延长的裁定。

你忠诚的

抄送:雇主

Dear Sir

We are in receipt of your instruction number [*insert number*] dated [*insert date*] instructing us to [*insert nature of work*] which you purport to issue under clause 3.13.1 of the conditions of contract. We consider such instruction unreasonable because [*state reasons*].

If within 7 days of receipt of this letter you do not in writing withdraw the instruction or modify it to remove our objection, a dispute or difference will exist as to whether the nature or extent of opening up/testing [*delete as appropriate*] in your instruction is reasonable in all the circumstances. Such dispute or difference will be referred to immediate adjudication/arbitration [*delete as appropriate*]. In such event, we will comply with our obligations pending the result of such adjudication/arbitration [*delete as appropriate*] and award of additional costs and extension of time.

Yours faithfully

Copy: Employer

函件 93
在隐蔽工程打开检查后,致建筑师

Letter 93
To architect, after work opened up for inspection

尊敬的先生:

  我方确认,遵照你方的指令把隐蔽工程[描述工程内容]打开供你方检查,你方于[填入日期]依据8.3条[当使用IFC98合同时,为"第3.12条";当使用MW98合同时,为"第3.5条";或当使用GC/Works/1(1998)合同时,为"第40(2)(i)条"]已检查了该工程,并且证实所用材料、货物和工程都是符合合同要求的。

  因此,打开及重新修复的费用应加进合同总价,过几天我方将把这些费用的细目提交给你方。我方将很快把详图、详细资料及预计超过竣工日的时间延误的计算,以及依据合同相应条款的赔偿申请提交给你方[当使用MW98合同时,代之以"雇主应考虑的赔偿"]。

你忠诚的

Dear Sir

We confirm that you inspected [*describe work*] opened up for your inspection in accordance with your instructions under clause 8.3 [*substitute* "*3.12*" *when using IFC 98*, "*3.5*" *when using MW 98 or* "*40(2) (i)*" *when using GC/ Works/ 1 (1998)*] on [*insert date*] and found the materials, goods and work to be in accordance with the contract.

The cost of opening up and making good, therefore, is to be added to the contract sum and we will let you have details of our costs within the next few days. We will shortly send you details, particulars and estimate of the expected delay in completion of the works beyond the completion date and an application for reimbursement under the appropriate clause of the contract [*substitute* "*reimbursement for consideration by the employer*" *when using MW 98*].

Yours faithfully

函件 94

当挖方工程可供检查时,致建筑师
本函只适用于 GC/Works/1(1998)合同

Letter 94

To architect, if excavations ready for inspection
This letter is only suitable for use with GC/Works/1 (1998)

尊敬的先生:

  本项目基础的挖方工程将于[填入日期]完成。
  为了避免延误,我方依据第 16 条规定,要求你方在该挖方工程完成时立即进行检查,并给予书面认可。

你忠诚的

Dear Sir

The excavations for foundations for this project will be complete by [*insert date*].

In order to avoid delay, we require you, in accordance with clause 16, to examine such excavations immediately on completion and give us your written approval.

Yours faithfully

函件 95
当指令拆除缺陷工程之后又签发了指令时,致建筑师
本函仅适用于 JCT98 合同或 WCD98 合同

Letter 95
To architect, if he issues an instruction after ordering removal of defective work
This letter is only suitable for use with JCT 98 or WCD 98

尊敬的先生:

  收到你方依据合同条件第 8.4.1 条于[填入日期]签发的关于把工程/材料/货物[视情况取舍]清除出工地现场的[填入编号]指令。现在我方又收到了一个依据第 8.4.3 条[当使用 WCD98 合同时,为"第 8.4.2 条"]于[填入日期]签发的要求变更[当使用 WCD98 合同时,代之以"改变"]的[填入号码]指令。

  [然后任选下列一段加入]

  一旦时机成熟,该工程将立即实施,当它完成时,我方将征得你方的认可。

  [或]

  你方依据第 8.4.1 条签发的指令,是没有必要的,因为[填入理由]。因此,要求你方重新考虑并立即撤销该指令。若你方通告我方,你方不准备撤销该指令,那么我方将会在时机成熟之时,立即执行你方的指令,但在无损害我方权利的前提下,我方将把此事提交仲裁解决。

你忠诚的

Dear Sir

We are in receipt of your instruction number [*insert number*] dated [*insert date*] under clause 8.4.1 of the conditions of contract regarding the removal from site of work/materials/goods [*delete as appropriate*]. We have now received a further instruction number [*insert number*] dated [*insert date*] requiring a variation [*substitute "change" when using WCD 98*] in consequence under clause 8.4.3 [*substitute "8.4.2" when using WCD 98*].

[*Then add either*:]

The work will be carried out as soon as reasonably practicable and we shall invite your approval when it is complete.

[*Or*:]

Your instruction is not reasonably necessary as a consequence of your clause 8.4.1 instruction because [*insert reasons*]. We request you to reconsider and to withdraw the instruction forthwith. If you inform us that you are not prepared to withdraw the instruction, we shall carry it out as soon as reasonably practicable, but without prejudice to our right, which we intend to exercise, to refer the matter to arbitration.

Yours faithfully

施工现场活动　　　　　　　　　　　　　　　　137

函件 96
当指令清除缺陷工程之后又签发要求打开隐蔽工程的指令时,致建筑师
本函仅适用于 JCT98 合同或 WCD98 合同

Letter 96
To architect, if he issues instruction for opening up after ordering removal of defective work
This letter is only suitable for use with JCT 98 or WCD 98

尊敬的先生:

　　收到你方依据合同条件第 8.4.1 条于[填入日期]签发的关于把工程/材料/货物[视情况取舍]清除出工地现场的[填入编号]的指令。现在我方又收到了一个依据第 8.4.4 条[当使用 WCD98 合同时,为"第 8.4.3 条"]于[填入日期]签发的要求打开隐蔽工程供检查的[填入编号]指令。

　　[然后加入下列任何一段]

　　该隐蔽工程将按要求于[填入日期][填入时间]打开供你方检查。

　　[或]

　　依我方的意见,你方忽视了实施合同的施工规程,而且,该指令并不合理。因此,我方建议把该指令看作是依据第 8.3 条的规定而签发的,若工程被证明是符合合同的,那么,我方期盼得到损失和(或)费用的合理补偿和工期顺延。请尽快复函确认,否则,在我们双方之间将会发生争端。我方将执行你方的指令,但不能剥夺我方把此事提交裁决或仲裁的权利。

　　　　　　　　　　　　　　　　　　　　　　　　　　　你忠诚的

Dear Sir

We are in receipt of your instruction number [*insert number*] dated [*insert date*] under clause 8.4.1 of the conditions of contract regarding the removal from site of work/materials/goods [*delete as appropriate*]. We have now received a further instruction number [*insert number*] dated [*insert date*] requiring opening up for inspection under clause 8.4.4 [*substitute "8.4.3" when using WCD 98*].

[*Then add either:*]

The work will be opened up as requested to be ready for your inspection on [*insert date*] at [*insert time*].

[*Or:*]

In our view, due regard has not been given to the code of practice in the contract and further, the instruction is not reasonable. We therefore propose to treat the instruction as being issued under the provisions of clause 8.3 and, if the work is found to be in accordance with the contract, we expect proper remuneration to include loss and/or expense and an appropriate extension of time. Please confirm your agreement by return or a dispute will have arisen between us and we shall carry out your instruction without prejudice to our right to refer the matter to adjudication or arbitration.

Yours faithfully

函件 97
若为拆除缺陷工程签发指令,致建筑师
本函只适用于 JCT98 合同

Letter 97
To architect, if issuing instruction for removal of defective work
This letter is only suitable for use with JCT 98

尊敬的先生:

  收到你方[填入日期]编号为[填入编号]的指令,要求把你方认为不符合合同的工程从工地现场拆除,你方认为该工程是[重述建筑师的描述]。
  我方的记载资料表明上述提及的工程是在[填入日期]约[填入一个时间段]天之前施工的。该工程实质上是建筑师依据第 2.1 条感到满意的工程。我方提请你方注意合同条件第 8.2.2 条,它规定你方必须在工程实施之后的一个合理时间内表明你方的任何不满意。很清楚,你方违反了合同,至今没有明确表达这样的不满意。现在我方有权要求为执行你方的指令而得到付款,若能尽快回函确认,我方将不胜感激。

你忠诚的

Dear Sir

We are in receipt of your instruction number [*insert number*] dated [*insert date*] instructing us to remove from site work which you state is not in accordance with the contract. You specify the work as [*repeat the architect's description*].

Our records show that the work in question was executed on [*insert date or dates*], some [*insert time period*] ago. The work was inherently a matter for the architect's satisfaction under clause 2.1. We draw your attention to clause 8.2.2 of the conditions of contract which stipulates that you must express any dissatisfaction with such work within a reasonable time from its execution. It is clear that, in breach of contract, you have not so expressed such dissatisfaction. We are now entitled to payment for complying with your instruction and we should be pleased to have your agreement by return.

Yours faithfully

函件 98
当工程将被隐蔽时,致建筑师
本函仅适用于 GC/Works/1(1998)合同

Letter 98
To architect, if work to be covered up
This letter is only suitable for use with GC/Works/1 (1998)

尊敬的先生:

　　依据合同条件第 17 条我方在此通知,[细述工程内容]将在[填入日期]被覆盖,在那天之后,任何要求打开所产生的费用将由雇主承担。

你忠诚的

Dear Sir

We hereby give notice, in accordance with clause 17 of the conditions of contract, that [*describe work*] will be covered up on [*insert date*]. After this date, any opening up required will be at the employer's expense.

Yours faithfully

函件 99
当工地现场发现文物时,致建筑师
本函不适用于 IFC98 合同或 MW98 合同

Letter 99
To architect, if antiquities found
This letter is not suitable for use with IFC 98 or MW 98

尊敬的先生:

  今天我方发现了我方认为是属于第 34 条 [当使用 GC/Works/1(1998) 合同时,为"第 3.2.3 条"] 规定的物品 [描述内容]。它位于 [描述位置,例如:检查井 23 号的西侧 2m 处],埋置深度为 [填入附近永久性的标高测定点] 下 [填入精确深度]。

  我方已停止了附近的工作,并做了临时木围栏和防水盖。我方希望立即得到你方的书面指令。

  [若使用 JCT98 合同或 WCD98 合同,加入:]

  工程进度由于上述情况遭到了延误,我方认为这是一个与第 25.4.5.1 条相关的事件。等该延误结束时,我方将向你方提交一份工期延误的计算和进一步的支持文件。你方尽可放心,我方正在竭尽全力防止工程进度和工程竣工的延误。请把本函作为依据第 25.2.1 条所发出的通知书。我方还认为依据第 34.3.1 条的规定,我方有权要求得到直接损失和(或)费用的补偿。

  [若使用 GC/Works/1(1998) 合同,加入:]

  依据第 36(1) 条,我方通知你方,很明显工程将不可能在竣工日竣工。在目前阶段,我方尚不能估算出总的延误天数,但当此事件结束时,我方将立即提交详细的测算细节。你方尽可放心,我方正在竭尽全力防止工程的延误以及把不可避免的延误减到最小程度。

  由于上述事件,工程的正常进度可能遭到了实实在在的耽搁和拖延。因此,为执行合同,我方将肯定多发生一些直接费用,而这些费用原本不会发生或超过了合同中所合理预见的费用。依据第 41 条,我方认为理应在合同总价中增加一笔费用。

你忠诚的

Dear Sir

We have today uncovered [*describe*] which we consider falls within the provisions of clause 34 [*substitute "32(3)" when using GC/Works/1 (1998)*]. It lies at a depth of [*insert precise depth*] below [*insert some nearby permanent point from which measurement taken*] at [*describe location e.g. "2 metres west of proposed inspection chamber 23"*].

We have stopped work in the vicinity and erected a temporary wooden fence and waterproof cover. We should be pleased to receive your immediate written instructions.

[*If using JCT 98 or WCD 98, add*:]

The progress of the works is being delayed by the above circumstances which we consider to be a relevant event under clause 25.4.5.1. When the delay is finished we will furnish you with our estimate of delay in the completion of the works and further supporting particulars. You may be assured that we are using our best endeavours to prevent delay in progress and completion of the works. Please take this as notice in accordance with clause 25.2.1. We further consider that we are entitled to reimbursement for direct loss and/or expense under the provisions of clause 34.3.1.

[*If using GC/Works/1 (1998), add*:]

In accordance with clause 36(1) we notify you that it is apparent that the works will not be complete by the date for completion. We are unable to estimate the total delay at this stage, but when it is over, we will submit full details immediately. You may be assured that we are using our best endeavours to prevent delays and to minimise unavoidable delays.

The regular progress of the works is likely to be materially disrupted and prolonged due to the above circumstances. In consequence, we will properly and directly incur expense in performing the contract which we would not otherwise have incurred and which is beyond that otherwise provided for in or reasonably contemplated by the contract. We expect to be entitled to an increase in the contract sum under clause 43.

Yours faithfully

第五章　工程款支付

　　大多数承包商都认为确保工程款支付是其工作中最棘手的问题之一。本书中所涉及的所有标准格式，除了 GC/Works/1(1998) 合同外，都为承包商提供了相应条款，当雇主未在规定期限内支付工程款（或在 WCD98 合同中支付所申请的金额）时，使得承包商能够决定其雇佣关系。如果仅仅是建筑师未在适当的时间签发付款证书，也是一种雇主应承担责任的违约行为。在这种情况下，没有付款证书也有可能获得应付款补偿。

　　更为困难的一种情形是，建筑师签证的金额少于承包商所申报的金额。如果说服无效，只能由裁决或仲裁解决争端。仲裁可能是一个漫长的过程，建筑师可能在进行仲裁之前就在随后的付款证书中纠正其错误。即使相对快速的裁决程序，从开始到结束，通常也要 35 天时间，这段时间足够建筑师签发另一个证书。

　　以下信函涉及了几种常见情况，包括建筑师未在规定时间签发最终付款证书的情形。还有一些专用函件适用于在 WCD98 合同中的特别付款情形。

函件 100a
致建筑师,附上期中付款申请
本函件仅适用于 WCD98 合同

Letter 100a
To architect, enclosing interim application for payment
This letter is only suitable for use with WCD 98

尊敬的先生:

　　根据第 30.3.1 条的规定,随信附上期中付款申请,我方应收金额系根据第 30.2A 条[当使用选项 B,为"第 30.2B 条"]计算。作为申请的依据材料,我方还附上第 30.3.2 条所要求的以下详细情况:

　　[列出详细清单]

　　我方提请你方注意第 30.3.6 条,该条款规定,你方必须在收到本申请的 14 天内支付本申请所指明的应付金额。

你忠诚的

Dear Sir

Under the provisions of clause 30.3.1, we enclose an application for interim payment stating the amount due to us calculated in accordance with clause 30.2A [*substitute "30.2B" if alternative B applies*]. In support of our application we enclose the following details as required under clause 30.3.2:

[*List*]

We draw your attention to clause 30.3.6 which stipulates that you must pay the amount stated as due in this application within 14 days of its receipt.

Yours faithfully

函件 100b
致建筑师,附上期中付款申请
本函件仅适用于 MW98 合同或 GC/Works/1(1998)合同

Letter 100b
To architect, enclosing interim application for payment
This letter is only suitable for use with MW 98 or GC/Works/1 (1998)

尊敬的先生:

  随信附上一份工程进度款[当使用 GC/Works/1(1998)合同时,代之以"即付预付款"]申请,我方应收金额系根据第 4.2 条[当使用 GC/Works/1(1998)合同时,为"第 48 条"]计算。作为申请的依据材料,我方还附上以下文件:

[列出文件清单]

你忠诚的

Dear Sir

We enclose an application for progress payment [*substitute "payment of advance on account" when using GC/Works/1 (1998)*] stating the amount due to us calculated in accordance with the provisions of clause 4.2 [*substitute "48" when using GC/Works/1 (1998)*]. In support of our application, we enclose the following documents:

[*List*]

Yours faithfully

函件 101
致工料测量师，提交评估申请
本函件仅适用于 JCT98 合同或 IFC98 合同

Letter 101
To quantity surveyor, submitting valuation application
This letter is only suitable for use with JCT 98 or IFC 98

尊敬的先生：

随信附上一份关于进行工程估价的申请。我方认为这些工程应在按第 30.2 条 [当使用 IFC98 合同时，为"第 4.2.1 条和第 4.2.2 条"] 计算的估价范围之内。

[在 JCT98 合同下，若指定的分包商已向承包商提出申请，加上：]

本申请包括了由 [填入相关指定分包商的姓名] 根据所订分包合同第 4.17 条提出的申请。

[然后加上：]

在此谨告，根据第 30.1.2.2 条 [当使用 IFC98 合同时，为"第 4.2(c) 条"]，为了确定期中付款证书中的应付款额，你方现在必须作出估价。对不同意我方申请的部分，你方必须向我方递交一份确认这些不同意见的说明。请注意：该说明必须与我方申请所含的内容同等详细。

你忠诚的

Dear Sir

We enclose an application setting out what we consider to be the valuation calculated in accordance with clause 30. 2 [*substitute* "*clauses 4. 2. 1 and 4. 2. 2*" *when using IFC 98*].

[*If, under JCT 98, a nominated sub-contractor has made an application to the contractor, add*:]

The application includes applications by [*insert the name(s) of the relevant nominated sub-contractors*] in accordance with clause 4. 17 of the nominated sub-contract.

[*Then add*:]

May we remind you that, under clause 30. 1. 2. 2 [*substitute* "*4. 2(c)*" *when using IFC 98*], you must now make a valuation for the purpose of ascertaining the amount due in an interim certificate and, to the extent that you disagree with our application, you must submit to us a statement which identifies such disagreement. Note that the statement must be in similar detail to that contained in our application.

Yours faithfully

函件 102
若工料测量师未对估价申请做出回应时,致建筑师
本函件仅适用于 JCT98 合同或 IFC98 合同

Letter 102
To architect, if quantity surveyor fails to respond to the valuation application
This letter is only suitable for use with JCT 98 or IFC 98

尊敬的先生:

  兹确认[填入日期]签发的、编号为[填入编号]付款证书业已收到。由该证书可见,确认的金额实际上少于我方根据第 30.1.2.2 条 [使用 IFC98 合同时, 为 "第 4.2(c)条"] 所申请的金额。因与该条款的明确规定相悖,工料测量师却未向我方提交一份与我方的申请同等详细的说明,来表明对我方申请金额的任何不同意见,我方对此十分关注。这本该应在工料测量师进行估价时,即不迟于下一付款证书签发前 7 天就应提交的。

  工料测量师的这一失职已构成违约,我方因此有权获得适当的赔偿。如若直至估价日仍未收到工料测量师详细的不同意见,我方就有理由认为工料测量师不持异议。我方因此并出于信任,承担了不菲的费用。现在我方意识到,这些费用将无法收回。

  我方认为,你方若能立即按两份估价之间的差额补签一份付款证书,可以避免赔偿问题。

你忠诚的

Dear Sir

We acknowledge receipt of certificate number [*insert number*] dated [*insert date*] from which we see that the amount valued was substantially less than the amount included in our application submitted under the provisions of clause 30.1.2.2 [*substitute "4.2(c)" when using IFC 98*]. We are concerned, because contrary to the express provisions of the clause, the quantity surveyor has not submitted to us a statement, in similar detail to our application, identifying any disagreement with our application. This should have been done at the time of the quantity surveyor's valuation, i.e. no later than 7 days before the date of the next certificate.

The quantity surveyor's failure amounts to a breach of contract for which we are entitled to appropriate damages. When we did not receive the quantity surveyor's detailed disagreement by the valuation date, we were entitled to assume that the quantity surveyor did not disagree. Accordingly and in reliance, we undertook expenditure of sums of money which we now realise we are not going to receive.

We suggest that damages could be avoided if you would immediately issue a supplementary certificate for the difference between the two valuations.

Yours faithfully

函件 103
若期中付款证书未签发,致建筑师
本函件不适用于 WCD98 合同

Letter 103
To architect, if interim certificate not issued
This letter is not suitable for use with WCD 98

尊敬的先生:

  我方迄今尚未收到期中付款证书的副本。根据合同条件第 30.1.1.1 条 [当使用 IFC98 合同时,为"第 4.2(a)条";当使用 MW98 合同时,为"第 4.2.1 条";当使用 GC/Works/1(1998)合同时,为"第 50(1)条"],该付款证书本应于[填入日期]签发。

  该付款证书可能在投递过程中遗失。若果真如此,要是你方能再发一份副本,我方将不胜感激。

  倘若事实上你方并未签发这一付款证书,我方必须提醒你方,签发这一付款证书是你方的合同责任,并应于[填入日期]前送达我方。否则,我方将立即对雇主违约采取法律诉讼行为。

你忠诚的

抄送:雇主

Dear Sir

We have not received a copy of the interim certificate which should have been issued on the [*insert date*] in accordance with clause 30.1.1.1 [*substitute* "4.2(a)" *when using IFC 98 or* "4.2.1" *when using MW 98 or* "50(1)" *when using GC/Works/1 (1998)*] of the conditions of contract.

It may be that the certificate has been lost in the post and on this assumption we should be pleased if you would send us a further copy.

If you have not, in fact, issued a certificate, we must remind you of your contractual duty so to do and request that it is in our hands by [*insert date*]. Failing which, we will take immediate legal action against the employer for the breach.

Yours faithfully

Copy: Employer

函件 104

若期中付款证书签证金额不足,致建筑师

本函件不适用于 WCD98 合同

Letter 104

To architect, if certificate insufficient

This letter is not suitable for use with WCD 98

尊敬的先生:

　　于[填入日期]签发、编号为[填入编号]的期中付款证书收悉。我方注意到[填入有争议的数字]与我方所提交文件中的证据、与你方/工料测量师[视情况取舍]的讨论或现场情况不相符。

　　倘若于 [填入日期] 未收到更正后的期中付款证书,我方将就此立即寻求裁决解决。

你忠诚的

抄送:雇主

Dear Sir

We have received your interim certificate number [*insert number*] dated [*insert date*]. We note [*insert the disputed figures*] which do not correspond with the evidence in the documents we have submitted, our discussions with you/the quantity surveyor [*delete as appropriate*] or the situation on site.

If we do not receive a corrected certificate by [*insert date*] we shall seek immediate adjudication on the matter.

Yours faithfully

Copy: Employer

函件 105
若未全额付款,且未发出拒付通知,致雇主
专递/挂号邮件

Letter 105
To employer, if payment not made in full and no withholding notice issued
Special / recorded delivery

尊敬的先生:

　　你方金额为[填入金额]的支票于今日收悉。但这一金额比建筑师于[填入日期]签发给我方的、编号为[填入编号]的付款证书中所确认的应付款金额短少[填入短少金额]。我方注意到,你方已经拒付了[填入金额],但并未对此给出任何理由/你方所送达的通知已经过时/你方所送达的通知并未对这一拒付给出正当的详细理由 [视情况取舍]。

　　因此,你方的行为已构成违约,并与《住宅转让、建造与改建法案1996》的有关条款规定相悖。倘若未于[填入日期]收到你方所拒付的[填入拒付金额]这一款额,我方将在我方认为合适时,实施本合同或普通法赋予的补救措施。

你忠诚的

抄送:建筑师

Dear Sir

We have today received your cheque for [*insert amount*], some [*insert difference*] less than certified due to us in the architect's certificate number [*insert number*] dated [*insert date*]. We note that you have withheld [*insert amount*], but you have given no reasons for doing so/ the notice you have sent is out of time/ the notice you have sent does not properly particularise the grounds for withholding [*delete as appropriate*].

Therefore, your action amounts to a breach of contract and it is contrary to the provisions of the Housing Grants, Construction and Regeneration Act 1996. If we do not receive the sum of [*insert the amount withheld*] by [*insert date*] we will exercise our contractual or common law remedies as we deem appropriate.

Yours faithfully

Copy: Architect

函件 106
若未在规定期限内支付预付款,致雇主
本函件仅适用于 JCT98 合同或 IFC98 合同

Letter 106
To employer, if the advance payment is not paid on the due date
This letter is only suitable for use with JCT 98 or IFC 98

尊敬的先生:

　　根据第 30.1.1.6 条[当使用 IFC98 合同时,为"第 4.2(b)条"],总额为[填入数额]的预付款应按合同附录规定,于[填入日期]支付给我方。但时间已逾[填入天数]天。我方于[填入日期]提供了一份保函,该保函的标准格式已经你方认可。

　　你方未支付预付款是严重的违约行为。我方有赖于该款项作为本项目的启动资金。事实上,这一款项的支付与否是我方作出是否签订本合同的重要因素。

　　我方认为,我方有权就你方的违约行为获得损害赔偿金。倘若几天后你方仍不支付预付款,由于这将严重影响我方工程的进程,我方可能认为你方拒付预付款。在此种情况下,我方将接受这一违约行为,并结束我方的合同义务。我方不希望如此,而是期盼收到该预付款。

你忠诚的

Dear Sir

Under the provisions of clause 30.1.1.6 [*substitute* "4.2(b)" *when using IFC 98*] an advance payment in the sum of [*insert the amount*] was due to be paid to us on the [*insert date*] as stated in the appendix. That is [*insert number*] days ago. We furnished a bond on the [*insert date*] in the standard terms from a surety which you have approved.

Your failure to provide the advance payment is a serious breach of contract. We were relying on the payment to assist our funding of this project. Indeed, the offer of such payment was an important factor in deciding to enter into this contract.

We are advised that we are entitled to damages for your breach and that, if it continues for more than a few days, we may be able to treat it as repudiatory, because it will effectively prevent us from proceeding with the works. In that case, we would be able to accept the breach and bring our obligations to an end. Hopefully, that will not be necessary and we look forward to receiving your payment by return.

Yours faithfully

函件 107

若评估未按标价的分项工程表进行时,致建筑师

本函件仅适用于 JCT98 合同或 IFC98 合同

Letter 107

To architect, if valuation not carried out in accordance with the priced activity schedule

This letter is only suitable for use with JCT 98 or IFC 98

尊敬的先生:

　　[填入日期]签发的、编号为[填入编号]的付款证书刚刚收悉。由该付款证书可见,所签证的金额实际上少于我方的期望金额。由此反映出评估并没有按照我方所提供的标价的分项工程表进行。

　　我方提请你方注意第 30.2.1.1 条[使用 IFC98 合同时,为"第 4.2.1(a)条"],该条款专门在附录中规定了附带标价的分项工程表,分项工程表有关的造价必须是各子项造价的总和,而各子项造价是将工程某一子项在某一分项工程中所占比例计算形成的。

　　现附上一个简明的分项计算,表明是通过这样的计算产生的数据。毫无疑问,(签证金额短少的)问题仅仅是因疏忽产生的。然而,鉴于签证金额的严重短缺,我方认为应立即再签发一份补充付款证书以纠正这一状况。

你忠诚的

Dear Sir

We have just received certificate number [*insert number*] dated [*insert date*] from which we see that the amount certified is substantially less than we expected. It appears that the valuation has not been carried out using the priced activity schedule we supplied.

May we draw your attention to clause 30.2.1.1 [*substitute "4.2.1(a)" when using IFC 98*] which specifically provides that where the appendix states that a priced activity schedule is attached, the value of work to which it relates must be the total of the various sums which result after the proportion of work in a particular activity is applied to the price for the work in the activity schedule.

A simple breakdown is enclosed showing the figure produced by applying that calculation. No doubt the matter is simply the result of an oversight. However, in view of the serious shortfall in the amount certified, we believe that a further and immediate supplementary certificate should be issued to rectify the position.

Yours faithfully

函件 108
致建筑师,要求为场外材料付款
本函件仅适用于 JCT98 合同或 IFC98 合同

Letter 108
To architect, requesting payment for off-site materials
This letter is only suitable for use with JCT 98 or IFC 98

尊敬的先生:

下列各项物品和材料为"在册物品",储存在[填入场所],你方随时可进行检查:

[列出详细清单]

若你方按合同条件第 30.3 条[当使用 IFC98 合同时,为"第 4.2.1(c)条"],若能在你方下一次期中付款额中将这些物品和材料的价值包括在内,我方将不胜感激。我方确认,就这些物品和材料,我方已满足了合同中的全部有关要求,由所附文件证明其为我方财产。

[视情况加上:]

我方确认,我方已按合同附件的格式提供了担保。

你忠诚的

Dear Sir

The following goods and materials are 'listed items' and are stored at [*insert place*] and available for your inspection at any time:

[*List*]

We should be pleased if you would operate the provisions of clause 30. 3 [*substitute "4. 2. 1 (c)" when using IFC 98*] of the conditions of contract and include the value of such goods and materials in your next interim valuation. We confirm that we have complied with all the requirements of the contract in respect of such goods and materials which the enclosed documents prove are our property.

[*Add, if appropriate*:]

We confirm that we have provided a bond in terms as annexed to the contract.

Yours faithfully

函件 109
致雇主,提前 7 天发出暂停施工的通知
专递邮件

Letter 109
To employer, giving 7 days notice of suspension
Special delivery

尊敬的先生:

我方注意到你方未在最终付款日支付/未能全额支付[视情况取舍]应付款。在我方的其他权利和赔偿不受损害的前提下,谨发出如下通知:除非你方在收到本函件后 7 天内支付全部的应付款额,否则我方将根据第 30.1.4 条[当使用 IFC98 合同时,为"第 4.4A 条";当使用 MW98 合同时,为"第 4.8 条";当使用 WCD98 合同时,为"第 52 条";当使用 GC/Works/1(1998)合同时,为"第 30.3.8 条"]的规定,在所有款项收讫前暂缓履行我方所有的合同义务。

你忠诚的

抄送:建筑师[仅当使用 JCT98 合同、IFC98 合同或 MW98 合同时]

Dear Sirs

We note that you have failed to pay/pay in full [*delete as appropriate*] the sum due by the final date for payment. Without prejudice to our other rights and remedies, we give notice that unless you pay the amount due in full within 7 days after receipt of this letter, we shall, in accordance with clause 30.1.4 [*substitute* "4.4A" *when using IFC 98*, "4.8" *when using MW 98*, "30.3.8" *when using WCD 98 or* "52" *when using GC/Works/1 (1998)*], suspend performance of all our obligations under the contract until full payment is received.

Yours faithfully

Copy: Architect [*only when using JCT 98, IFC 98 or MW 98*]

函件 110
若无视暂停施工的通知，7天内仍未全额付款时，致雇主

Letter 110
To employer, if payment in full has not been made within 7 days despite notice of suspension

---

尊敬的先生：

  我方在[填入日期]的函中指出，若在该函收到后7天内未全额付清应付款，将暂缓履行我方的合同义务。但此后我方仍未收到/全额收到[视情况取舍]付款。作为直接后果，现谨通知你方，我方暂缓履行我方所有的合同义务。

  我方已使现场处于安全状态，但若你方坚持拒付，则由你方负责现场的安全和保险。

你忠诚的

Dear Sir

Further to our letter dated [*insert date*] stating that if payment in full was not made within 7 days from the date of its receipt we would suspend our obligations, we have not received any/full [*delete as appropriate*] payment. This is to inform you that, with immediate effect, we are suspending all our obligations under the contract.

We have left the site safe, but it is now for you to arrange security and insurance if you persist in withholding payment.

Yours faithfully

函件 111
致雇主,要求支付拖欠款项的利息
本函件不适用于 GC/Works/1(1998)合同

Letter 111
To employer, requesting interest on late payment
This letter is not suitable for use with GC/Works/1（1998）

尊敬的先生:

  就[填入日期]签发的、编号为[填入编号]的付款证书,最终付款日应为[填入日期]。迄至发函时止,我方尚未收到任何/全部[视情况取舍]付款。

  按第30.1.1.1条[当使用IFC98合同时,为"第4.2(a)条";当使用MW98合同时,为"第4.2.2条";当使用WCD98合同时,为"第30.3.7条"]的规定,你方必须向我方支付在应付款拖欠之日起至款额结清之日止的利息,利率按应付款拖欠之日当日的英国银行基准利率上浮5%计算。然而,你方应注意要求支付利息的这一合同权利不影响我方其他的权利和赔偿。尽管若在收到本函后次日下班前尚未付款,我方仍会来函要求另行补偿,但本函在此通知,我方无意放弃要求支付利息的权利。

你忠诚的

Dear Sir

We refer to certificate number [*insert number*] dated [*insert date*]. The final date for payment was [*insert date*]. At the time of writing we have not received any/full [*delete as appropriate*] payment.

Under the provisions of clause 30.1.1.1 [*substitute*"4.2(a)" *when using IFC 98,* "4.2.2" *when using MW 98 and* "30.3.7" *when using WCD 98*] you are obliged to pay us simple interest at 5% above Bank of England Base Rate current at the date payment became overdue until the amount is paid. However, you should note that this contractual right to interest is without prejudice to our other rights and remedies. This letter is simply by way of notice that we have no intention of waiving our right to interest although we intend to write to you under separate cover if payment is not made by close of business on the day following the date of this letter.

Yours faithfully

函件 112
致雇主,要求把保留金存入独立的银行账户之内
本函件不适用于 MW 98 或 GC/Works/1(1998)合同
专递/挂号邮件

Letter 112
To employer, requesting retention money to be placed in a separate bank account
This letter is not suitable for use with MW 98 or GC/Works/1 (1998)
Special/recorded delivery

尊敬的先生:

　　我方正式要求你方将所有现有和以后的保留金存入专设的独立的银行账户,并将所存款额认同是为我方的利益而托管。请告知我方此账户的开户银行名、账户名和账号。

　　尽管合同有规定,但你方的义务无疑不因我方可能提出的任何正式要求而改变。

你忠诚的

抄送:建筑师

Dear Sir

We formally request you to place all current and future retention money in a separate bank account set up for the express purpose and identified as money held in trust for our benefit. Please inform us of the name of the bank, the account name and number.

Notwithstanding the provisions of the contract, it is established that your obligation exists irrespective of any formal request we may make.

Yours faithfully

Copy: Architect

函件 113

若未将保留金存入独立的银行账户,致雇主

本函件不适用于 MW 98 或 GC/Works/1(1998)合同

专递/挂号邮件

Letter 113

To employer, if failure to place retention in separate bank account

This letter is not suitable for use with MW 98 or GC/Works/1 (1998)

Special/recorded delivery

尊敬的先生:

　　就我方[填入日期]的函,你方未将保留金存入到一个我方所要求的、也是合同和法律所规定的独立信托基金账户。

　　独立信托基金将会在你方破产时保护我方的资金。倘若你方未能满足我方于[填入日期]所提出的要求,我方将立即寻求一个强迫使你方服从的指令。

你忠诚的

抄送:建筑师

Dear Sir

Further to our letter dated [*insert date*] you have not notified us that you have set aside retention money in a separate trust fund as we requested and as the contract and the law provides.

The separate trust fund will protect our money in the event of your insolvency. If you have not complied with our request by [*insert date*] we shall immediately seek an injunction to compel you to comply.

Yours faithfully

Copy: Architect

函件 114
致建筑师,附上为准备最终付款证书所需的所有资料
本函件不适用 WCD98 合同或 GC/Works/1(1998)合同

Letter 114
To architect, enclosing all information for preparation of final certificate
This letter is not suitable for use with WCD 98 or GC/Works/1 (1998)

尊敬的先生:

　　根据第 30.6.1.1 条[当使用 IFC98 合同时,为"第 4.5 条";当使用 MW98 合同时,为"第 4.5.1.1 条"],谨附上本合同最终账单的细节,以及所有的依据文件。
　　倘若你方进行必要的计算和验算,以使最终付款证书能够按本合同的时间表签发,我方将不胜感激。

你忠诚的

Dear Sir

In accordance with clause 30.6.1.1 [*substitute "4.5" when using IFC 98 or "4.5.1.1" when using MW 98*] we enclose full details of the final account for this contract together with all supporting documentation.

We should be pleased if you would proceed with the necessary calculations and verifications to enable the final certificate to be issued in accordance with the contract timescale.

Yours faithfully

函件 115
致建筑师,附上最终账单
本函件仅适用 WCD98 合同

Letter 115
To architect, enclosing final account
This letter is only suitable for use with WCD 98

尊敬的先生:

根据合同条件第 30.5.1 条,谨附上此最终账单,以及与第 30.5.4 条有关的最终报表,如蒙同意,我方将不胜感激。

并附上以下依据文件:

[列出文件清单]

倘若你方要求进一步资料,我方将十分乐意提供。请在收到本函后一个月内给我方一份清单,完整列出此种要求(如有)。

你忠诚的

Dear Sir

In accordance with clause 30.5.1 of the conditions of contract we enclose the final account and final statement referred to in clause 30.5.4 and we should be pleased to have your agreement.

We enclose the following supporting documentation:

[*List*]

If you reasonably require any further information, we will be pleased to provide it. Please let us have a complete list of such requirements (if any) within one month from the date of this letter.

Yours faithfully

函件 116a

若未在预定日期签发最终付款证书,致建筑师

本函件仅适用于 JCT98 合同

专递/挂号邮件

Letter 116a

To architect, if final certificate not issued on due date

This letter is only suitable for use with JCT 98

Special/recorded delivery

尊敬的先生:

　　合同条件第 30.8 条要求你方自以下各项事件中的最迟的事件起 2 个月内签发最终付款证书:

　　1. 缺陷责任期结束——[填入日期]。
　　2. 缺陷整改完成证书签发日——[填入证书上的日期]。
　　3. 你方送交我方根据第 30.6.1.2 条的核定书和说明的副本的日期——[填入日期或,若尚未送交核定书和说明,增加以下内容:]在本事件中,你方尚未送交我方这样一个应在[填入日期]签发的核定书和说明。

　　因此,最终付款证书应于[填入按以上计算的日期]签发。自此时间已逾[填入周数]周,而我方尚未收到这一付款证书。你方已违约,我方认为,雇主应对此负责。但倘若最终付款证书于[填入日期]送达,在我方对此权利不受损害的前提下,我方将不会对这一违约行为采取进一步的行动。

你忠诚的

抄送:雇主

Dear Sir

Clause 30.8 of the conditions of contract requires you to issue the final certificate within 2 months from the latest of the following events:

1. The end of the defects liability period — [*insert date*].

2. Date of issue of certificate of completion of making good defects — [*insert date on certificate*].

3. The date on which you sent us a copy of the ascertainment and statement under clause 30.6.1.2 — [*insert date or, if the architect has not sent an ascertainment and statement, add the following*:] In this instance, you have not sent us such ascertainment and statement which should have been issued on the [*insert date*].

Therefore, the final certificate should have been issued on the [*insert date calculated as above*]. Some [*insert number*] weeks have passed since that date and we have received no such certificate. You are in breach of contract, a breach for which, we are advised, the employer is liable. Without prejudice to our rights in this matter, if the final certificate is in our hands by [*insert date*], we will take no further action on such breach.

Yours faithfully

Copy: Employer

函件 116b
若未在预定日期签发最终付款证书,致建筑师
本函件仅适用于 IFC98 合同
专递/挂号邮件

Letter 116b
To architect, if final certificate not issued on due date
This letter is only suitable for use with IFC 98
Special/recorded delivery

尊敬的先生:

合同条件第 4.6.1.1 条要求你方自以下各项事件中最迟的事件起 28 天内签发最终付款证书:

1. 向我方送递的经调整的合同总价的计算书,我方收讫日期为[填入日期]。
2. 你方按第 2.10 条的证书签发日——[填入日期]。

因此,最终付款证书应于[填入按以上计算的日期]签发。自此时间已逾[填入周数]周,而我方尚未收到这一付款证书。你方已违约,我方认为,雇主应对此负责。但倘若最终付款证书于[填入日期]送达,在我方对此权利不受损害的前提下,我方将不会对这一违约行为采取进一步的行动。

你忠诚的

抄送:雇主

Dear Sir

Clause 4.6.1.1 of the conditions of contract requires you to issue the final certificate within 28 days of the latest of the following events:

1. The sending to us of the computations of the adjusted contract sum, which we received on the [*insert date*].
2. Your certificate under clause 2.10 —[*insert date*].

Therefore, the final certificate should have been issued on the [*insert date calculated as above*]. Some [*insert number*] weeks have passed since that date and we have received no such certificate. You are in breach of contract, a breach for which, we are advised, the employer is liable. Without prejudice to our rights in this matter, if the final certificate is in our hands by [*insert date*], we will take no further action on such breach.

Yours faithfully

Copy: Employer

函件 116c
若未在预定日期签发最终付款证书,致建筑师
本函件只适用于 MW98 合同
专递/挂号邮件

Letter 116c
To architect, if final certificate not issued on due date
This letter is only suitable for use with MW 98
Special / recorded delivery

尊敬的先生:

按合同条件第 4.4.1.1 条要求,只要你方已签发了第 2.5 条规定的付款证书,就应在收到对需核实的工程量计算的所有依据文件后的 28 天内签发最终付款证书。

你方在[填入日期]的来函中确认,已收到所有这些文件。你方于[填入日期]签发了第 2.5 条规定的付款证书。

因此,最终付款证书应于[填入按以上计算的日期]签发。自此时间已逾[填入周数]周,而我方尚未收到这一付款证书。你方已违约,我方认为,雇主应对此负责。但倘若最终付款证书于[填入日期]送达,在我方对此权利不受损害的前提下,我方将不会对这一违约行为采取进一步的行动。

你忠诚的

抄送:雇主

Dear Sir

Clause 4.4.1.1 of the conditions of contract requires you to issue the final certificate within 28 days of receipt by you of all documentation reasonably required for computation of the amount to be certified, provided that you have issued your certificate under clause 2.5.

You confirmed that all such documents were in your possession by your letter of the [*insert date*]. You issued a clause 2.5 certificate on the [*insert date*].

Therefore, the final certificate should have been issued on the [*insert date calculated as above*]. Some [*insert number*] weeks have passed since that date and we have received no such certificate. You are in breach of contract, a breach for which, we are advised, the employer is liable. Without prejudice to our rights in this matter, if the final certificate is in our hands by [*insert date*], we will take no further action on such breach.

Yours faithfully
Copy: Employer

第六章　工期顺延

对工期的延误和顺延的全部事务存在很大的误解。合同中工期顺延条款存在的主要原因是可使已知的竣工日期固定,以符合雇主负责或雇主应承担风险的事件带来的工期延误。这样工期延误的损失可以得到弥补。次要的理由是在双方无法控制的引起工期延误的事件中,承担部分承包商的风险。除 MW98 合同以外的所有标准合同格式,一旦发生工期延误对可能发生的情况,都有详细的规定。为了确保在某一特定合同下有权获得的工期顺延,必须十分小心地遵循关于工期延误通知和提供依据材料的详细规定。虽然工期顺延申请不是批准工期顺延的先决条件,但是不正确遵守这些规定,常常造成工期顺延批准不及时,并且工期顺延的时间会比期望的要短得多。

以下的函件涉及工期延误通知,以及通常因建筑师或承包商不十分了解各自的义务而产生的问题的处理。

函件 117
若出现工期延误,但又没有工期顺延的理由时,致建筑师
本函件不适用于 MW98 合同或 GC/Works/1(1998)合同

Letter 117
To architect, if delay occurs, but no grounds for extension of time
This letter is not suitable for use with MW 98 or GC/Works/1 (1998)

尊敬的先生:

  工程进度因[陈述原因]已经/看来要[视情况取舍]延误。

  我方会继续尽全力使这一延误及其影响减到最小程度,并且会立即向你方报告这一延误的原因。

  本通知依据第 25.2.1.1 条[当使用 IFC98 合同时,为"第 2.3 条"]发出。

你忠诚的

Dear Sir

The progress of the works is being/is likely to be [*delete as appropriate*] delayed due to [*state reasons*].

We will continue to use our best endeavours to minimise the delay and its effects and we will inform you immediately the cause of the delay has ceased to operate.

This notice is issued in accordance with clause 25.2.1.1 [*substitute* "2.3" *when using IFC 98*].

Yours faithfully

函件 118

若造成工期延误的事件已停止,但又没有理由顺延工期时,致建筑师

本函件不适用于 MW98 合同或 GC/Works/1(1998)合同

Letter 118

To architect, when cause of delay ended if no grounds for extension of time

This letter is not suitable for use with MW 98 or GC/ Works/ 1 (1998)

尊敬的先生:

  本函为我方于[填入日期]向你方发函有关工期延误一事。

  我方谨向你方通告,这一工期延误的事件已经处理。在工期延误期间我方所采取的措施,以及对以后几周的后续工作所作出的计划,能够确保赶上失去的时间,并使工程进度符合[填入日期]认可的工程进度计划。

你忠诚的

Dear Sir

We refer to our letter of the [*insert date*] notifying you of delay.

We are pleased to be able to inform you that the cause of the delay has been dealt with. The measures which we adopted during the period of delay and the continuing procedures over the next few weeks are designed to enable us to recover the lost time and put the progress of the works on programme by [*insert date*].

Yours faithfully

函件 119a
若出现工期延误,并有工期顺延的理由时,致建筑师
本函件不适用于 MW98 合同或 GC/Works/1(1998)合同

Letter 119a
To architect, if delay occurs giving grounds for extension of time
This letter is not suitable for use with MW 98 or GC/Works/1 (1998)

尊敬的先生:

  工程进度因[陈述原因]已经延误。我方认为这是与[填入序号]条款相关的事件。

  这一延误起始于[填入日期]。当这一延误结束时,我方会呈交我方对工程竣工延误的估算和进一步的详细材料。

  你方可以确信,我方正在尽全力防止工程进度延误和竣工的推迟。

  本通知发出的依据是合同条件第 25.2.1.1 条 [当使用 IFC98 合同时,为"第 2.3 条"]。

你忠诚的

抄送:指定的分包商[仅适用于使用 JCT98 合同,并在原通知中提及时]

Dear Sir

The progress of the works is being delayed due to [*state reasons*]. We consider this to be a relevant event under clause [*insert number*].

The delay began on [*insert date*]. When it is finished, we will furnish you with our estimate of delay in the completion of the works and further supporting particulars.

You may be assured that we are using our best endeavours to prevent delay in progress and completion of the works.

This notice is issued in accordance with clause 25.2.1.1 [*substitute "2.3" when using IFC 98*].

Yours faithfully

Copy: Nominated sub-contractor [*applies only when using JCT 98 and if reference made in original notice*]

函件 119b
若出现工期延误,并有工期顺延的理由时,致建筑师
本函件仅适用于MW98合同或GC/Works/1(1998)合同

Letter 119b
To architect, if delay occurs giving grounds for extension of time
This letter is only suitable for use with MW 98 or GC/Works/1 (1998)

尊敬的先生:

已经很明显,工程将不能在竣工日按时竣工。

[然后可以:]

情况是[陈述原因]。我方估计这一延误的总时间为[填入时间段]。我方认为我方有权得到工期的顺延[填入时间段],并且会在以后的几天内向你方呈交进一步的详细依据材料。

[也可以:]

情况是[陈述原因]。目前我方尚不能估算这一延误的总时间,因为问题还在继续中。当延误结束时,我方会立即提供所要求的估算和其他细节。

[然后:]

你方可以确信,我方正在尽全力防止工程进度延误和竣工的推迟。
发出本通知的依据是合同条件第2.2条[当使用GC/Works/1(1998)合同时,为"第36(1)条"]。

你忠诚的

Dear Sir

It is apparent that the works will not be complete by the date for completion.

[*Then either*:]

The circumstances are [*state*]. We estimate the delay to total [*insert period*]. We consider that we are entitled to an extension of [*insert period*] and we will furnish further particulars within the next few days.

[*Or*:]

The circumstances are [*state*]. We are unable to estimate the total delay at this stage because it is continuing. When the delay is over, we will submit the required estimate and other details immediately.

[*Then*:]

You may be assured that we are using our best endeavours to prevent delay in progress and completion of the works.

This notice is issued in accordance with clause 2.2 [*substitute* " *36(1)* " *when using GC/Works/1 (1998)*] of the conditions of contract.

Yours faithfully

函件 120a
致建筑师,提供工期顺延所需的进一步详细材料
本函件不适用于 MW98 合同或 GC/Works/1(1998)合同

Letter 120a
To architect, providing further particulars for extension of time
This letter is not suitable for use with MW 98 or GC/Works/1 (1998)

尊敬的先生:

就我方于[填入日期]的函,告知工程进度的延误可能要引起工程竣工的推迟。以下,我方就每一有关事项,告知这一延迟对预计工程竣工的影响程度:

[分别列出有关事项,估计每种情况下工程竣工的推迟时间,以及提供相关依据材料]我方相信,你方现在已有充分的根据,来批准一个公正合理的工期顺延时间。

你忠诚的

抄送:指定的分包商[仅适用于使用JCT98合同,并在原通知中提及时]

Dear Sir

We refer to our letter of the [*insert date*] in which we notified you of a delay in progress of the works likely to result in a delay to completion of the works. We note below particulars of the expected effects and the estimated extent of delay in completion of the works in respect of each relevant event specified in our notice:

[*List relevant events separately, giving an assessment of the delay to completion in each case together with any other supporting information*]

We believe that you now have sufficient information to enable you to grant a fair and reasonable extension of time.

Yours faithfully

Copy: Nominated sub-contractor [*applies only when using JCT 98 and if reference made in original notice*]

函件 120b

致建筑师,提供工期顺延所需的进一步详细材料

本函件仅适用于 MW98 合同或 GC/Works/1(1998)合同

Letter 120b

To architect, providing further particulars for extension of time

This letter is only suitable for use with MW 98 or GC/Works/1 (1998)

尊敬的先生:

就我方于[填入日期]的通知,告知工程进度的延误可能要引起工程竣工的推迟。

我方估计竣工将要推迟[填入周数]周,同时认为,我方有权获准工期顺延。我方是根据以下的理由得出此结论的:

[全面陈述理由,并包括所有的依据材料]。

我方相信,你方现在已有充分的根据,来批准一个公正合理的工期顺延时间。切盼回音。

你忠诚的

Dear Sir

We refer to our notice of the [*insert date*] in which we informed you of a delay in progress of the works likely to result in a delay to completion of the works.

We estimate that completion will be delayed by [*insert number*] weeks and we consider that we should be granted an extension of time for that period. We arrive at this conclusion as follows:

[*state reasons in fall and include all supporting information*].

We believe that you now have sufficient information to enable you to grant a fair and reasonable extension of time and we look forward to hearing from you shortly.

Yours faithfully

函件 121
当建筑师要求提供进一步详细材料以批准工期的顺延,致建筑师

Letter 121
To architect, if he requests further information in order to grant extension of time

---

尊敬的先生:

　　对你方[填入日期]的来函,深表谢意。来函要求提供关于[陈述建筑师的要求]的进一步资料。

　　[以对建筑师的提问做确切回答的格式陈述其所要求的资料,或者,若建筑师未说明其要求时,则按如下书写:]

　　我方相信,我方已在[填入日期]的函件中给予你方所要求的所有资料。不过从你方来函中尚不清楚现在究竟要求提供何种资料。要是你方能提出具体问题,我方当竭诚答复,并提供所能考虑到的进一步的详细材料。
　　切盼回音。

你忠诚的

Dear Sir

Thank you for your letter of the [*insert date*] in which you request further information in respect of [*state what architect requires*].

[*State the information required by the architect in the form of precise answers to his questions or, if the architect does not say what he requires, write as follows*:]

We believe that we gave you all the information you require in our letter of the [*insert date*]. It is not clear from your letter what further information you now request. If you would be good enough to ask specific questions, we will do our best to answer them and supply whatever further supporting details then suggest themselves to us.

We look forward to hearing from you as soon as possible.

Yours faithfully

函件 122
当建筑师不合理地要求提供进一步详细材料以批准工期的顺延，致建筑师

Letter 122
To architect, if he unreasonably requests further information in order to grant an extension of time

尊敬的先生：

  对你方于[填入日期]的来函，要求提供关于[陈述建筑师的要求]的进一步资料以批准工期的顺延，深表谢意。

  我方已按合同要求于［填入日期］提交了工期延误通知［若使用 GC/Works/1(1998)合同，代之以"工期顺延请求"］。我方提交了全部详细材料，包括这一延误对[填入日期，或者视情况代之以"在同一时间"]竣工日的影响。我方相信，你方现在已有全部必要根据，来公正合理地批准将工期顺延至[填入日期]。当然，尽快向你方提供足够的资料符合我方的利益，我方已经做到了这一点。

  依我方之见，你方最近要求我方提供材料，只不过是拖延工期顺延的批准时间而已。因而我方正式请求你方按第 25.3.1 条[若使用 IFC98 合同时，为"第 2.3 条"；若使用 MW98 合同时，为"第 2.2 条"；若使用 GC/Works/1(1998)合同，为"第 36(1)条"]履行责任。

你忠诚的

Dear Sir

Thank you for your letter of the [*insert date*] requesting further information in order to enable you to grant an extension of time.

We submitted notice of delay [*substitute "request for extension of time" when using GC/Works/1 (1998)*] as required by the contract, on [*insert date*]. We submitted full particulars including estimate of the effect of the delay on completion date on [*insert date or dates or substitute "at the same time" if appropriate*]. We believe that you had all the information necessary to enable you to make a fair and reasonable extension of time by [*insert date*]. It is, of course, very much in our interests to supply you with full information as quickly as possible; this we have done.

It is our view that your latest request for information is nothing but an attempt to postpone the granting of an extension. We, therefore, formally call upon you to carry out your duty under clause 25.3.1 [*substitute "2.3" when using IFC 98, "2.2" when using MW 98 or "36 (1)" when using GC/Works/1 (1998)*].

Yours faithfully

函件 123
若建筑师所批准的工期顺延时间不够,致建筑师

Letter 123
To architect, if extension of time is insufficient

尊敬的先生:

来函今日收悉,得知工期已顺延[填入时间段],新的竣工日期为[填入日期]。

我方发现你方的结论与事实和我方所提交的依据材料不符。

或许你方能重新考虑所批准的工期顺延时间,或者向我方说明你方批准上述工期顺延时间的理由。

你忠诚的

Dear Sir

We have received today your notification of an extension of time of [*insert period*] producing a new date for completion of [*insert date*].

We find your conclusions inexplicable in the light of the facts and the information we submitted in support of those facts.

Perhaps you would be good enough to reconsider your grant of extension of time or let us have an indication of your reasons for arriving at the time period you have granted.

Yours faithfully

函件 124
若建筑师所批准的工期顺延时间不够,并且不打算重新考虑时,致建筑师

Letter 124
To architect, if extension of time is insufficient and he is not willing to reconsider

---

尊敬的先生：

　　对你方[填入日期]的来函,深表谢意。在我方[填入日期]发出通知,并按你方[填入日期]的要求于[填入日期]提交了进一步的资料后,注意到你方仍不打算重新考虑所批准的工期顺延时间。

　　[若建筑师给出了理由：]

　　我方已经仔细地研究了你方所作决定的理由,发现你方忽视了我方提交文件的许多内容以及该问题的许多事实。

　　[若建筑师未给出理由：]

　　我方注意到你方拒绝对你方的决定给出任何理由。我方只能认为你方的依据并不可靠。

　　[然后加上：]

　　倘若你方认为全面讨论是有益的,我方将十分乐意与你方会晤。否则,我方会将这一争端立即提交裁决/仲裁解决[视情况取舍]。

你忠诚的

Dear Sir

Thank you for your letter of the [*insert date*] from which we note that you are not willing to reconsider your grant of extension of time in response to our notice of the [*insert date*] and submissions of further information of the [*insert date*] in response to your request of the [*insert date*].

[*If the architect has given reasons*:]

We have carefully examined the reasons you give in support of your decision and they reveal that you have ignored much of our submission and the facts of the matter.

[*If the architect has not given reasons*:]

We note that you refuse to give any reasons for your decision and we can only assume that you are on uncertain ground.

[*Then add*:]

We will be happy to meet you if you think that a full discussion would be helpful. Failing that, we intend to refer this dispute to immediate adjudication/arbitration [*delete as appropriate*].

Yours faithfully

函件 125
若未在规定时间内批准工期顺延,致建筑师
本函件不适用于 IFC98 合同或 MW98 合同

Letter 125
To architect, if extension of time not granted within time stipulated
This letter is not suitable for use with IFC 98 or MW 98

尊敬的先生:

  工期延误通知[当使用 GC/Works/1(1998)合同时,代之以"工期顺延请求"]已按合同条件第 25.2.1.1 条[当使用 GC/Works/1(1998)合同时,为"第 36(1)条"]于[填入日期]送交你方。包括这一延误对竣工影响的估算和所要求的对工期顺延的估算在内的全部详细材料已于[填入日期]送交你方。你方并未要求进一步的材料。第 25.3.1 条[当使用 GC/Works/1(1998)合同时,为"第 36(1)条"]要求你方在收到详细材料后的不迟于 12 周[当使用 GC/Works/1(1998)合同时,代之以"第 42 天"]内将决定通知我方。这一期限已于[填入日期]期满,但你方并未将决定通知我方。显然你方已违约,雇主应对此承担责任。

  你方没有授权就目前的有关事件[当使用 GC/Works/1(1998)合同时,代之以"情况"]批准在合同竣工日[当使用 GC/Works/1(1998)合同时,代之以"工程竣工"]的顺延。由于你方违约,我方可能承受的无论是原材料增加或其他造成的任何损失和开支,均应在适当时候由雇主赔付。

你忠诚的

抄送:雇主

Dear Sir

Notice of delay [*substitute "Request for extension of time" when using GC/Works/1 (1998)*] was sent to you on [*insert date*] in accordance with clause 25.2.1.1 [*substitute "36(1)" when using GC/Works/1 (1998)*] of the conditions of contract. Full particulars including estimate of the delay to completion and estimate of the extension required were sent to you on [*insert date*]. You made no request for further information. Clause 25.3.1 [*substitute "36(1)" when using GC/Works/1 (1998)*] requires you to notify us of your decision not later than 12 weeks [*substitute "42 days" when using GC/Works/1 (1998)*] from receipt of particulars [*substitute "notice" when using GC/Works/1 (1998)*]. The period elapsed on [*insert date*] and you have not informed us of your decision. You are clearly in breach of contract, a breach for which the employer is responsible.

You are not empowered to make any extension of time for the current relevant events [*substitute "circumstances" when using GC/Works/1 (1998)*] until after the contract completion date [*substitute "completion of the works" when using GC/Works/1 (1998)*]. Any loss or expense which we may suffer, whether from increasing resources or otherwise, as a result of your breach will be recovered from the employer as damages in due course.

Yours faithfully

Copy: Employer

函件 126
若建筑师拖延工期顺延的批准,致建筑师
本函件仅适用于 IFC98 合同或 MW98 合同

Letter 126
To architect, if slow in granting extension of time
This letter is only suitable for use with IFC 98 or MW 98

尊敬的先生:

　　工期延误通知已按合同条件第 2.3 条 [当使用 MW98 合同时,为"第 2.2 条"] 于 [填入日期] 送交你方。全部详细材料已于 [填入日期] 送交你方。

　　你方有 [填入周数] 周的时间来作出决定,现在要求你方批准我方应得到的工期顺延。倘若于 [填入日期] 仍未收到你方关于批准这一工期顺延的通知,你方就构成违约,雇主应对此承担责任。

　　[若使用 IFC98 合同,就加上:]

　　你方无权在工程实际竣工日后再对现时事件来决定工期的顺延。由于你方违约,我方可能承受的无论是原材料增加或其他造成的任何损失和开支,均应在适当时候由雇主赔付。

　　[若使用 MW98 合同,就加上:]

　　我方认为工期是"不确定"的,雇主将失去扣减工期延误损失费权力,因为竣工日期未固定,无法计算工期延误损失费用,你方也已失去了决定竣工日期的权力。我方的义务则是在一个合理的时间内完成工程的施工。

　　　　　　　　　　　　　　　　　　　　　　　　　　　　你忠诚的

抄送:雇主

Dear Sir

Notice of delay was sent to you on [*insert date*] in accordance with clause 2.3 [*substitute "2.2" when using MW 98*] of the conditions of contract. Full particulars were sent to you on [*insert date*].

You have now had [*insert number*] weeks in which to make your decision and we now call upon you to grant us the extension of time to which we are entitled. If we do not receive your notice granting such extension by [*insert date*] you will be in breach of contract, a breach for which the employer will be responsible.

[*If using IFC 98, add*:]

You will not be empowered to make any extension of time for the current events until after the date of practical completion. Any loss or expense which we may suffer, whether from increasing resources or otherwise, as a result of your breach will be recovered from the employer as damages in due course.

[*If using MW 98, add*:]

We will consider that time is "at large" and the employer will have lost his right to deduct liquidated damages because there will be no date for completion from which liquidated damages can run and you will have lost your power to fix such a date. Our obligation will then be to finish the works within a reasonable time.

Yours faithfully

Copy: Employer

函件 127a
若建筑师尚未审查工期顺延申请,致建筑师
本函件仅适用于 JCT98 合同或 WCD98 合同

Letter 127a
To architect, if review of extensions not carried out
This letter is only suitable for use with JCT 98 or WCD 98

尊敬的先生:

  合同条件第 25.3.3 条要求你方应当:

1. 确定一个迟于先前确定的日期的竣工日,或
2. 确定一个先于先前确定的日期的竣工日,或
3. 确认先前已确定过的竣工日。

  你方必须至少在实际竣工日之后 12 周内履行这一职责。无论是否存在相反的看法,我方认为,这一时间期限是合同义务。这一期限于 [填入日期] 期满,因而我方现在的义务仅仅是在一个合理时间内完成施工。因为竣工日期未定,无法计算工期延误损失,你方也无权来决定竣工日期,因此,雇主将失去扣减工期延误损失赔付费用的权利。任何扣减这一工期延误损失赔付费用的企图将会立即引起我方的法律诉讼。

你忠诚的

抄送:雇主

Dear Sir

Clause 25.3.3 requires you to either:

1. Fix a completion date later than that previously fixed, or

2. Fix a completion date earlier than that previously fixed, or

3. Confirm the completion date previously fixed.

You must carry out this duty, at latest, within 12 weeks after the date of practical completion. We are advised that, despite speculation to the contrary, the time period is mandatory. That period expired on [*insert date*] and our obligation now is simply to complete within a reasonable time. The employer has lost his right to deduct liquidated damages, because there is no date fixed for completion from which such damages can run and you have lost your power to fix such a date. Any attempt to deduct such damages will result in immediate legal action on our part.

Yours faithfully

Copy: Employer

函件 127b
若建筑师尚未审查工期顺延申请,致建筑师
本函件仅适用于 IFC98 合同

Letter 127b
To architect, if review of extensions not carried out
This letter is only suitable for use with IFC 98

尊敬的先生:

  不论你方是否审查过先前的决定,还是我方是否发过工期延误通知,合同条件第 2.3 条允许你方给予工期顺延。工程竣工后对工期进行审查,是你方的一项很重要的权力。

  你方必须至少在实际竣工日之后 12 周内履行这一职责。无论是否存在相反的看法,我方认为,这一时间期限是合同义务。这一期限于[填入日期]期满,因而我方现在的义务仅仅是在一个合理时间内完成施工。因为竣工日期未定,无法计算工期延误损失,你方也无权来决定竣工日期,因此,雇主将失去扣减工期延误损失赔付费用的权利。任何扣减这一工期延误损失赔付费用的企图将会立即引起我方的法律诉讼。

你忠诚的

抄送:雇主

Dear Sir

Clause 2.3 permits you to extend time whether upon reviewing previous decisions or otherwise and whether or not we have given notice of delay. This is a valuable power for you to review the situation after the works are finished.

However, you must carry out this duty within 12 weeks after the date of practical completion. We are advised that, despite speculation to the contrary, that time period is mandatory. That period expired on [*insert date*] and our obligation now is simply to complete within a reasonable time. The employer has lost his right to deduct liquidated damages, because there is no date fixed from which such damages can run and you have lost your power to fix such a date. Any attempt to deduct such damages will result in immediate legal action on our part.

Yours faithfully

Copy: Employer

函件 127c
若建筑师对工期顺延申请无最终决定时,致建筑师
本函件仅适用于 GC/Works/1(1998)合同

Letter 127c
To architect, if no final decision on extensions of time
This letter is only suitable for use with GC/Works/1（1998）

尊敬的先生:

　　合同条件第 36(4)条规定,在工程竣工后的 42 天内,你方必须就所有未定的和临时的工期顺延申请作出最终决定。无论是否存在相反的看法,我方认为,这一时间期限是合同义务。这一期限于[填入日期]期满,而你方未对原来于[填入日期]所通知的工期顺延的申请作出最终决定,因而我方现在的义务仅仅是在一个合理时间内完成施工。因为竣工日期未定,无法计算工期延误损失,你方也无权来决定竣工日期,因此,雇主将失去扣减工期延误损失赔付费用的权利。任何扣减这一工期延误损失赔付费用的企图将会立即引起我方的法律诉讼。

你忠诚的

抄送:雇主

Dear Sir

Clause 36(4) of the conditions of contract stipulates that you must come to a final decision on all outstanding and interim extensions of time within 42 days after completion of the works. We are advised that, despite speculation to the contrary, the time period is mandatory. The period expired on [*insert date*] and you have not made a final decision on the request(s) for extension of time originally notified to you on [*insert date or dates*] and our obligation now is simply to complete within a reasonable time. The employer has lost the right to deduct liquidated damages, because there is no date fixed for completion and you have lost your power to fix such a date. Any attempt to deduct such damages will result in immediate legal action on our part.

Yours faithfully

Copy: Employer

函件 128
若建筑师认为承包商未尽力而为时,致建筑师
本函件不适用于 MW98 合同

Letter 128
To architect, if he alleges that contractor is not using best endeavours
This letter is not suitable for use with MW 98

尊敬的先生:

  合同条件第 25.3.4.1 条 [当使用 IFC98 合同时,为"第 2.3 条";当使用 GC/Works/1(1998)合同时,为"第 36(6)条"]要求我方尽力而为以防止工程延误。你方在[指出地点和时间,例如,2002 年 9 月 3 日举行的现场会议记录第 7.4 条]中认为我方在这方面未履行合同职责是完全没有根据的。我方的合同义务仅仅是尽可能重新安排劳动力、继续有序而勤奋地施工。我方已经做到这一点,并且还将继续去做。但我方没有义务增加支出以弥补损失的工期。事若果真如此,工期顺延的条款就是不必要的了。

  倘若你方认为我方未尽力而为,因此要减少我方应得的工期顺延时间,我方会立即采取我方能采取的适当的补救对策。

你忠诚的

抄送:雇主

Dear Sir

Clause 25.3.4.1 [*substitute "2.3" when using IFC 98 or "36(6)" when using GC/Works/1 (1998)*] of the conditions of contract requires us to use our best endeavours to prevent delay. Your allegation in [*state where and date, e.g: minute no. 7.4 of the site meeting held on the 3 September 2002*] that we are failing to carry out our duties in this respect is totally without foundation. Our obligation to use best endeavours is simply an obligation to continue to work regularly and diligently, rearranging our labour force as best we can. This we have done and we are continuing so to do. There is no obligation upon us to expend additional sums of money to make up lost time. If that was the case, the extension of time clause would be otiose.

If you purport to reduce, on grounds of failure to use best endeavours, our entitlement to an extension of time, we will take immediate and appropriate advice on the remedies available to us.

Yours faithfully

函件 129
若雇主错扣了工期延误损失赔偿费用,致雇主

Letter 129
To employer, if he wrongfully deducts liquidated damages

尊敬的先生:

你方金额为[填入金额]的支票于今日收讫,但此金额比编号为[填入编号]的付款证书中应付给我方的金额短少[填入金额]。我方从你方[填入日期]来函中注意到,这一短缺的金额是因所谓的工期延误[填入周数]周而扣减我方应承担的[填入金额]的损失赔偿费。

我方认为,因为[陈述原因]的缘故,你方已违约。倘若未在[填入日期]收到金额为[填入数字]的支票,我方会采取适当的措施,不仅弥补错扣的金额,而且要弥补损失、利息及由此而发生的开支。我方保留按第28.2.1.1条[当使用IFC98合同时,为"第7.9.1(a)条";当使用MW98合同时,为"第7.3.1.1条"]拥有的解除雇佣关系的权利。

你忠诚的

Dear Sir

We have received your cheque today in the sum of [*insert amount*] which falls short of the amount certified as due to us in certificate number [*insert number*] by [*insert amount*]. We note from your accompanying letter of the [*insert date*] that the deficit represents your deduction of alleged liquidated damages in the sum of [*insert amount*] for [*insert number*] weeks.

We consider you to be in breach of contract because [*state reasons*]. If we do not receive your cheque for the full amount of [*insert amount*] by [*insert date*], we will take appropriate steps to recover not only the amount wrongfully deducted, but also damages, interest and costs. We reserve the right to determine our employment under clause 28.2.1.1 [*substitute "7.9.1(a)" when using IFC 98 or "7.3.1.1" when using MW 98. Omit this sentence when using GC/Works/1 (1998)*].

Yours faithfully

函件 130
若退回损失赔款时未含利息,致雇主

Letter 130
To employer, if damages repaid without interest

尊敬的先生:

  你方[填入日期]来函收悉,其中包括金额为[填入金额]的支票。这一金额是指错扣的/收回的工期延误损失赔偿费[视情况取舍]。我方在[填入日期]来函中提醒你方因你方违约造成损失应作出赔付的通知。我方准备接受以损失赔付额为基础的利息,并且这一额外付款是以按应付款拖欠当日的英格兰银行现行基准利率上浮5%乘以拖欠的天数计算得出的[填入金额以及计算过程]。

  我方认为这是由于你方无意的疏忽造成的。但在保留我方的一切权利和赔偿要求的前提下,倘若于[填入日期]收到金额为[填入利息金额]的款项,我方将不打算采取诉诸法律的任何行动。

你忠诚的

抄送:建筑师

Dear Sir

We are in receipt of your letter of the [*insert date*] enclosing your cheque for [*insert amount*] representing liquidated damages wrongfully deducted/recovered [*delete as appropriate*]. You will recall that in our letter dated [*insert date*] we gave you due notice that we should require damages for your breach. We are prepared to accept interest charges as a realistic basis for damages and the additional payment required on account of such interest is, therefore, calculated on the basis of 5% above Bank of England Base Rate current at the date payment became overdue [*insert amount together with the calculation*].

We assume that this is a genuine oversight on your part and, while reserving all our rights and remedies, we do not propose to take any action if we receive the sum of [*insert amount of interest*] by [*insert date*].

Yours faithfully
Copy: Architect

第七章　损失和(或)费用

如果意欲索赔损失和(或)费用，必须遵循合同规定的程序。否则，索赔会遭到拒绝。承包商有时没有认识到，建筑师对雇主承担的责任是不拒绝索赔，但只接受承包商能提供确凿证据的索赔。

最为重要的是在最佳时间提出申请。然而，记住在任何标准格式合同中都是获得工期顺延成为直接损失和(或)费用开支的支付的先决条件。这已经为法庭所确认。

如果不能满足合同条款，即使如果能满足，也可以按普通法为损失要求索赔。本书中作为参考的标准格式并不授予建筑师按普通法处理索赔的权力。

以下函件涉及最初的申请和提供进一步资料的条款。通常的难题是认定索赔的时间和索赔金额。在核定索赔金额前，对损失和(或)费用有确凿证据的承包商一直处于类似某人的现金被剥夺的地位。后面也包含了一些如何处理这类情况的建议性函件。

当然，如果建筑师对承包商所认为的应付损失和(或)费用不予签证，承包商可以将问题提交给快速处理的裁决程序，本书在后面会涉及这一点。

函件 131a

致建筑师,申请获得损失和(或)费用的付款

本函件不适用于 MW98 合同或 GC/Works/1(1998)合同

Letter 131a

To architect, applying for payment of loss and/or expense

This letter is not suitable for use with MW 98 or GC/Works/1 (1998)

尊敬的先生:

　　按合同条件第 26.1 条[当使用 IFC98 合同时,为"第 4.11 条"],谨提出如下申请:
　　我方已经承受/看来要承受[视情况取舍]直接损失和(或)费用以及为合同履约承担金融费用,却得不到根据本合同的任一其他条款的偿还付款,因为常规的工程进度实质上已经/看来要[视情况取舍]受[描述事件,除非涉及占有权延误,并加上:]的实际影响,成为第[填入条款序号]条下的问题。

你忠诚的

Dear Sir

We hereby make application under clause 26.1 [*substitute "4.11" when using IFC 98*] of the conditions of contract as follows:

We have incurred/are likely to incur [*delete as appropriate*] direct loss and/or expense and financing charges in the execution of this contract for which we will not be reimbursed by a payment under any other provision in this contract, because the regular progress of the works has been/is likely to be [*delete as appropriate*] materially affected by [*describe and unless referring to deferment of possession, add:*] being a matter in clause [*insert clause number*].

Yours faithfully

函件 131b
致建筑师,根据补充条款,申请获得损失和(或)费用的付款
本函件仅适用于 WCD98 合同
专递邮件

Letter 131b
To architect, applying for payment of loss and/or expense under the supplementary provisions
This letter is only suitable for use with WCD 98
Special delivery

尊敬的先生:

  根据合同条件第 26.1 条,我方有权得到一笔合同金额之外的补偿损失和(或)费用(包括金融费用)的金额。现附上根据合同条件第 30.3.1 条的应付款项的申请。因而根据补充条款第 S7.2 条,谨提交在申请之前的期间所承受的此种损失和(或)费用的估算,我方要求将其计入合同总价。只要我方继续发生直接损失和(或)费用,我方将继续按第 S7.3 条提交估算。

  请允许提醒你方在收到本估算的 21 天内务必给我方书面通知,以表明接受我方的估算,或希望协商,或按第 26 条的规定处理。

你忠诚的

Dear Sir

We are entitled to an amount in respect of loss and/or expense (including financing charges) to be added to the contract sum in accordance with clause 26.1 of the conditions of contract. Our application for payment under clause 30.3.1 is attached. Therefore in accordance with supplementary provision S7.2, we submit our estimate of such loss and/or expense, incurred in the period immediately preceding that for which such application is made, which we require to be added to the contract sum. We shall continue to submit estimates in accordance with clause S7.3 for so long as we continue to incur direct loss and/or expense.

May we remind you that within 21 days of receipt of this estimate you must give us written notice either that you accept our estimate, or that you wish to negotiate, or clause 26 will apply.

Yours faithfully

函件 131c
致建筑师,申请获得损失和(或)费用的付款
本函件仅适用于 MW98 合同

Letter 131c
To architect, applying for payment of loss and/or expense
This letter is only suitable for use with MW 98

尊敬的先生:

  谨通知你方,正常的工程进度实质上已经中断/拖延[视情况取舍],雇主或作为其代理人的你方应对此负责。依我方之见,这一问题超越了合同条件第3.6条预期的情况。该问题为[具体说明]。

  虽然合同似乎没有处理此类索赔的规定,我方还是可能按普通法而采取行动。不过我方相信,这一问题可以简单地由雇主或你方使用正常的权力得到解决。我方将愿意在此基础上处理这一问题,同时也乐意知道雇主对此想法是否一致。倘若雇主并无此意,或我方至[填入日期]尚未收到你方的回音,我方将按普通法正式提出索赔。

你忠诚的

抄送:雇主

Dear Sir

We hereby give notice that regular progress of the works has been materially disrupted/prolonged [*delete as appropriate*] by matters for which the employer or you as his agent are responsible. In our view, such matters exceed the situation contemplated by clause 3.6 of the conditions of contract. The matters are [*describe*].

Although the contract appears to have no machinery for dealing with claims of this nature, we are advised that we may bring an action at common law. We believe that the matters are capable of easy resolution by the employer or by you with proper authority. We should be content to proceed on this basis and we should be pleased to hear whether the employer is in agreement. If the employer is not prepared to deal with us on this basis or if we do not hear from you by [*insert date*] we shall formulate our claim at common law.

Yours faithfully

Copy: Employer

函件 131d

致建筑师,为费用申请补偿款

本函件仅适用于 GC/Works/1(1998)合同

Letter 131d

To architect, applying for payment of expense

This letter is only suitable for use with GC/Works/1 (1998)

尊敬的先生:

  按合同条件第 46 条第(3)(a)款,我方谨通知如下:

  正常的工程进度因为 [具体说明] 已经/看来要 [视情况取舍] 中断/拖延 [视情况取舍]。这一中断/拖延[视情况取舍]的后果,使我方在履行合同中完全并直接地蒙受了/将完全并直接地蒙受[视情况取舍]那些不该发生的费用。这些费用已经超越本合同中的内容或为合同所合理预见的内容。我方有权/期望有权 [视情况取舍] 按第 46(1)条增加合同总价。

你忠诚的

Dear Sir

We hereby give notice under clause 46(3)(a) of the conditions of contract as follows:
The regular progress of the works has been/is likely to be [*delete as appropriate*] disrupted/prolonged [*delete as appropriate*] due to [*describe*]. In consequence of such disruption/prolongation [*delete as appropriate*] we have properly and directly incurred/we will properly and directly incur [*delete as appropriate*] expense in performing the contract which we would not otherwise have incurred and which is beyond that otherwise provided for in or reasonably to be contemplated by the contract. We are/we expect to be [*delete as appropriate*] entitled to an increase in the contract sum under clause 46(1).

Yours faithfully

函件 132
致建筑师,提供有关损失和(或)费用的进一步细节
本函件不适用于MW98合同或GC/Works/1(1998)合同

Letter 132
To architect, giving further details of loss and/or expense
This letter is not suitable for use with MW 98 or GC/Works/1 (1998)

尊敬的先生:

对你方[填入日期]来函,深表谢意。来函要求再提供细节,作为对我方于[填入日期]提交的损失和(或)费用申请补偿款的依据。

谨附上一份以工序图/网络分析图[视情况取舍]形式表示的进度计划的副本。我方已作了标注以详细表明问题所在。可以从中看出[详细说明问题所在,给出日期和时间,以及操作技工人数和主要员工的姓名]。

随函还附上以下现场日记和工地会议纪录摘要的复印件:

[按日期列表。当使用WCD98合同不要包括以下段落]

倘若在你方按合同条件第26.1条[当使用IFC98合同时,为"第4.11条"]要求得出结论前,要求提供具体问题的更详细资料,请告知。我方将竭诚提供。

你忠诚的

Dear Sir

Thank you for your letter of the [*insert date*] requesting further information in support of our application for loss and/or expense which was submitted to you on [*insert date*].

We enclose a copy of the programme in precedence diagram/network analysis [*delete as appropriate*] form, which we have marked up to show the circumstances in some detail. It can be seen that [*describe the circumstances in some detail, giving dates and times, numbers of operatives and names of key members of staff involved*].
Also enclosed is the following copy correspondence, extracts from the site diary and site minutes:

[*List with dates. Do not include the following paragraph when using WCD 98*]

We should be pleased if you would inform us if there are any particular points on which you require more information before you are able to form an opinion as required by clause 26.1 [*substitute "4.11" when using IFC 98*] of the conditions of contract.

Yours faithfully

函件 133

致建筑师或工料测量师,附上损失和(或)费用的详细资料

本函件仅适用于 JCT98 合同或 IFC98 合同

Letter 133

To architect or quantity surveyor, enclosing details of loss and/or expense

This letter is only suitable for use with JCT 98 or IFC 98

尊敬的先生:

对你方[填入日期]来函,深表谢意。来函要求提供关于在我方 [填入日期]函中提及的详细资料。

谨附上该详情材料以及依据文件。倘若你方能按第 26.1 条 [当使用 IFC98 合同时,为"第 4.11 条"] 的要求进行损失和(或)费用补偿款的金额确定,我方将不胜感激。倘若你方还需进一步资料,请即告知。

你忠诚的

Dear Sir

Thank you for your letter of the [*insert date*] requesting details of loss and/or expense in respect of the matters notified in our letter of [*insert date*].

We enclose the details together with supporting documentation. We should be pleased if you would proceed with the ascertainment of loss and/or expense as required by clause 26.1 [*substitute* "4.11" *when using IFC 98*]. Please inform us immediately if you require any further information.

Yours faithfully

函件 134

致工料测量师,提供用于计算费用的资料

本函件仅适用于 GC/Works/1(1998)合同

Letter 134

To quantity surveyor, providing information for calculation of expense

This letter is only suitable for use with GC/Works/1 (1998)

尊敬的先生:

我方很高兴附上下列文件。这些资料反映了所发生的全部费用的细节,并证明了这些费用直接因发生了第 46(1)条中所描述的事件之一而产生。

切盼你方根据第 46(5)条,在收到本函后的 28 天内作出决定。

[列出文件和资料]

你忠诚的

Dear Sir

We have pleasure in enclosing the documents listed below. They contain full details of all expenses incurred and evidence that the expenses directly result from the occurrence of one of the events described in clause 46(1).

We look forward to your decision in accordance with clause 46(5) within 28 days of receipt of this letter.

[*List documents and information*]

Yours faithfully

函件 135a

若金额核定拖延,致建筑师
本函件仅适用于 JCT98 合同

Letter 135a

To architect, if ascertainment delayed
This letter is only suitable for use with JCT 98

尊敬的先生:

　　本函为我方根据合同条件第 26.1 条于 [填入日期] 提交的通知以及我方于 [填入日期] 的来函。该函附上了你方所要求的进一步资料,以得出结论并开展对损失和(或)费用金额的核定。

　　自上次向你方提交此种资料以来,时间已逾 [填入周数] 周。在此期间,你方或工料测量师均未要求我方再提交进一步的资料和详情。第 3 条明确规定了你方的责任,即在核定后的下一次期中付款证书中签发所核定的全部或部分金额。第 26.1 条指出,"一旦"你方认为工程受到影响,就应该"在此后不时地"去核定金额。第 26.5 条也有类似的规定。一旦已经得出结论,你方必须开始金额核定程序。一旦核定了金额,你方必须将其包括在下一次期中付款证书中。请将经你方核定并包括在下一次期中付款证书中的金额于 [填入日期] 通知我方。

你忠诚的

Dear Sir

We refer to our notice of the [*insert date*] submitted under the provisions of clause 26.1 of the conditions of contract and our letters of the [*insert dates*] enclosing the further information you required in order to form an opinion and carry out ascertainment of the amount of loss and/or expense.

[*Insert number*] weeks have elapsed since we last submitted such information to you and, during that period, neither you nor the quantity surveyor have requested further information or details. Clause 3 imposes a clear duty on you to certify sums ascertained in whole or in part in the next interim certificate following ascertainment. Clause 26.1 states that "as soon as" you are of the opinion that the works are affected, you will "from time to time thereafter" ascertain. A similar provision is contained in clause 26.5. As soon as you have formed your opinion, you must begin the process of ascertainment. As soon as any sum has been ascertained, you must include the amount in the next interim certificate. Please inform us by [*insert date*] the amount which you have ascertained and intend to include in the next certificate.

Yours faithfully

函件 135b
若金额核定拖延,致建筑师
本函件仅适用于 IFC98 合同

Letter 135b
To architect, if ascertainment delayed
This letter is only suitable for use with IFC 98

尊敬的先生:

　　本函为我方根据合同条件第 4.11 条于 [填入日期] 提交的通知以及我方于 [填入日期] 的来函。该函附上了你方所要求的进一步资料,以得出结论并开展对损失和(或)费用金额的核定。

　　自上次向你方提交此种资料以来,时间已逾 [填入周数] 周。在此期间,你方或工料测量师均未要求我方再提交进一步的资料和详情。第 4.2.2 条明确规定了你方的责任,即按第 4.11 条确定经核定的金额,在期中付款证书中签署"该金额已核定"。请在 [填入日期] 将经你方核定的金额通知我方,并计入下一次期中付款证书中。

你忠诚的

Dear Sir

We refer to our notice of the [*insert date*] submitted under the provisions of clause 4.11 of the conditions of contract and our letters of the [*insert dates*] enclosing the further information you required to form an opinion and carry out ascertainment of the amount of loss and/or expense.

[*Insert number*] weeks have elapsed since we last submitted such information to you and, during that period, neither you nor the quantity surveyor have requested further information or details. Clause 4.2.2 imposes a duty on you to certify ascertainments under clause 4.11 in your certification of interim payments "to the extent that such amounts have been ascertained". Please inform us by [*insert date*] the amount which you have ascertained and intend to include in the next certificate.

Yours faithfully

函件 136
若核定金额太少,致建筑师
本函件不适用于 WCD98 合同或 MW98 合同

Letter 136
To architect, if ascertainment too small
This letter is not suitable for use with WCD 98 or MW 98

尊敬的先生:

  本函为我方于[填入日期]关于损失和(或)费用[当使用 GC/Works/1(1998)合同时,代之以"费用"]的通知。你方[当使用 GC/Works/1(1998)合同时,代之以"工料测量师"]于[填入日期]的来函中,通知我方所核定的金额看来完全没有考虑我方于[填入日期]提交的依据极为充足的材料。

  除非于[填入日期]得知你方将修改核定的金额以考虑我方所提供的资料,否则将会引起争端,我方在适当时间将争端提交一种争端解决方案去解决。

你忠诚的

Dear Sir

We refer to our notice dated [*insert date*] in respect of loss and/or expense [*substitute "expense" when using GC/Works/1 (1998)*]. Your [*substitute "The quantity surveyor's" when using GC/Works/1 (1998)*] letter of the [*insert date*] notifying us of the amount ascertained appears to take little account of the very full supporting information submitted by us on [*insert date or dates*].

Unless we hear from you by [*insert date*] that you will amend your ascertainment to take account of the information we have supplied, a dispute will have arisen which we will refer to one of the dispute resolution procedures in due course.

Yours faithfully

函件 137
关于依据普通法的索赔,致雇主

Letter 137
To employer, regarding a common law claim

尊敬的先生:

现提请你方注意[描述引起索赔的情况以及日期]。这些情况使我方有权向你方提出损失索赔。

本合同对此种情况下的索赔并无规定,我方正在着手通过仲裁或诉讼的途径来获得损失的补偿。

你忠诚的

抄送:建筑师

Dear Sir

We draw your attention to [*describe the circumstances giving rise to the claim with dates*]. These circumstances entitle us to a claim for damages against you.

The contract makes no provision for such a claim in such circumstances and we are proceeding with a view to recovering damages by way of arbitration or litigation.

Yours faithfully

Copy: Architect

函件 138
关于依据普通法的索赔,致雇主
毫无偏见

Letter 138
To employer, regarding a common law claim
WITHOUT PREJUDICE

尊敬的先生:

　　就我方于[填入日期]函中通知的情况,兹援引普通法向你方提出索赔。
　　倘若你方准备与我方会晤,讨论索赔事宜,以达到问题的合理解决,我方将不采取直接的法律程序。
　　若蒙同意,请于[填入日期]通知我方,并请告知我方会晤可能举行的日期。

你忠诚的

抄送:建筑师

Dear Sir

We refer to the claim we have against you at common law arising from the circumstances notified to you in our letter of the [*insert date*].

If you are prepared to meet us to discuss our claim with a view to reaching a reasonable settlement of the matter, we will take no immediate legal proceedings.

Please inform us by [*insert date*] if you agree to this suggestion and let us know a date when such a meeting could take place.

Yours faithfully

Copy: Architect

第八章 合同终止、仲裁、裁决和工程竣工

合同终止是一项非常严肃的程序，它会带来许多危险。其中，最为重要的情形之一就是，如果没有认真遵循合同终止程序，就可能要为不按法律履行合同及由此给雇主带来的损害承担责任。合同终止通知必须按照一定的方式和格式送递给合同规定的当事人。本章的信函标准格式都涉及这类情况。但是在实际发出合同终止通知之前，应该谨慎地征求专家的意见。

仲裁同样是一项非常严肃的程序，且成本很高。当要开始考虑仲裁时，应听取有关法律和合约部门的建议。本章中还包括一些寻求共同聘请仲裁人的标准格式函件。

1996年住宅转让、建造和改建法案[1997年在北爱尔兰的（北爱尔兰）施工合同法令]规定了施工合同中的裁决程序，本章中收集了一些这方面有用的函件。裁决程序正广泛用于处理那些从未处理过的极为复杂且风险很大的争端事件。当有可能涉及这类争端或涉及那些需要法律支持的争端时，请确保获得相应的支持。一般来讲，雇主不大可能对承包商启动裁决程序，尽管偶尔也会发生，通常都是承包商启动裁决程序。

本章中的其他函件涉及工程竣工和缺陷责任期的内容。

函件 139 当发出违约通知时,致雇主或建筑师 本函不适用于 MW98 或 GC/Works/1(1998)合同 专递/挂号邮件 **Letter 139** To employer or architect, if default notice served *This letter is not suitable for use with MW 98 or GC/Works/1 (1998)* *Special/recorded delivery*
尊敬的先生: 你方[填入日期]的来函收悉。显而易见,你方准备根据合同条件第 27.2 条[使用 IFC98 合同时,为"第 7.2.1 条"]规定把该函视为违约通知。 [或添入以下内容:] 你方所谓的违约通知有严重的错误。 [或添入以下内容:] 我方对你方所谓的违约通知有疑义。 [或添入以下内容:] 你方所谓的违约通知中提到的违约内容是不正确的。 [或添入以下内容:] 你方所谓的违约通知提到的违约是你方违约的结果[视情形加以解释]。 [然后,再填入以下内容] 因此,你方的违约通知是无效的。请注意,如果你方随后发出合同终止通知,这将是不合法的,是一项违约行为,我方将立即对你方/雇主[视情形取舍]采取诉讼程序。   你忠诚的 抄送:建筑师/雇主[视情形取舍]

Dear Sir

We are in receipt of your letter of the [*insert date*], which apparently you intend to be a default notice in accordance with clause 27.2 [*substitute "7.2.1" when using IFC 98*] of the conditions of contract.

[*Add either*:]

The purported notice contains serious error.

[*Or*:]

We are advised that your purported notice is ambiguous.

[*Or*:]

The substance of the default specified in your purported notice is incorrect.

[*Or*:]

The default specified in your purported notice is the result of your own default. [*Explain as appropriate*].

[*Then add*:]

Your notice is, therefore, invalid and of no effect. Take notice that if you proceed to give notice of determination, it will be unlawful, it will amount to repudiation and we will take immediate proceedings against you/the employer [*delete as appropriate*].

Yours faithfully
Copy: Architect/employer [*delete as appropriate*]

函件 140

当发出合理的违约通知时,致雇主或建筑师
本函不适用于 MW98 或 GC/Works/1(1998)合同
专递/挂号邮件

Letter 140

To employer or architect, if default notice served justly
This letter is not suitable for use with MW 98 or GC/Works/1 (1998)
Special/recorded delivery

尊敬的先生:

  你方于[填入日期]依据合同条件第27.2条[使用IFC98合同时,为"第7.2.1条"]规定签发的违约通知收悉。

  对你方认为有必要发出这样的违约通知,我方感到遗憾。现告知你方[填入将采取消除违约行为的措施]

你忠诚的

抄送:建筑师/雇主[视情形取舍]

Dear Sir

We are in receipt of your letter of the [*insert date*] which you sent as a default notice in accordance with clause 27.2 [*substitute "7.2.1" when using IFC 98*] of the conditions of contract.

We regret that you have felt it necessary to send such a notice, but we are pleased to be able to inform you that [*insert whatever steps are being taken to remove the default*].

Yours faithfully

Copy: Architect/employer [*delete as appropriate*]

函件 141
如果已发出不成熟的合同终止通知函时,致雇主
本函不适用于 MW98 或 GC/Works/1(1998)合同
专递/挂号邮件

Letter 141
To employer, if premature determination notice issued
This letter is not suitable for use with MW 98 or GC/Works/1 (1998)
Special/recorded delivery

尊敬的先生:

你方[填入日期]依据合同条件第27.2.2条[使用IFC98合同时,为"第7.2.2条"]规定发出的终止与我方雇佣关系的函收悉。

你方原先发出的违约通知函是[填入日期]收到的。邮局将会向你方确认违约通知函邮递的日期。因此,你方发出的合同终止通知是不成熟的,因此是无效的。对此,我方将对遭受的实质性的损害进行索赔。我方已纠正了违约通知中提及的违约行为,因此,在不影响我方任何权利及赔偿要求的情况下,尤其是(但无任何限制)在我方把你方所谓的合同终止视为违约行为加以对待的权利无任何影响的情况下,我方将按正常情况继续工作,同时愿接受合适的建议。

你忠诚的

Dear Sir

We are in receipt of your notice dated [*insert date*] purporting to determine our employment under clause 27.2.2 [*substitute "7.2.2" when using IFC 98*] of the conditions of contract.

Your original default notice was received on the [*insert date*]. The Post Office will be able to confirm to you the date of delivery. Your notice of determination was, therefore, premature and of no effect and may amount to a repudiation of the contract for which we can claim substantial damages. We have already corrected the default specified in your original notice and, without prejudice to any of our rights and remedies in this matter and particularly (but without limitation) our right to treat your purported determination as repudiation, we will continue to work normally while we take appropriate advice.

Yours faithfully

函件 142

致雇主,合同终止前发出违约通知
本函不适用于 GC/Works/1(1998)合同
专递/挂号邮件

Letter 142

To employer, giving notice of default before determination
This letter is not suitable for use with GC/Works/1 (1998)
Special/recorded delivery

尊敬的先生:

　　依据合同条件第 28.2.1/28.2.2 条 [视情形取舍,使用 IFC98 合同时,为"第 7.9.1/7.9.2 条";使用 MW98 合同时,为"第 7.3 条"] 规定,现正式通知你方:你方已违约/发生了特定的停工事件[视情形取舍],具体如下:

　　[视情形说明违约的细节及发生的日期并参阅相应的合同条文]

　　上述情形必须结束。
　　若你方在收到我方的违约通知函后 14 天 [使用 MW98 合同时,为 7 天] 里继续违约/发生了特定的停工事件 [视情形取舍],我方将依据本合同立即终止我们之间的雇佣关系,恕不再告知。

你忠诚的

Dear Sir

We hereby give you notice under the provisions of clause 28.2.1/28.2.2 [*delete as appropriate or substitute* "7.9.1/7.9.2" *when using IFC 98 or* "7.3" *when using MW 98*] of the conditions of contract that you are in default/a specified suspension event has occurred [*delete as appropriate*] in the following respect:

[*Insert details of the default with dates if appropriate and refer to the appropriate sub-clause*]

which must be ended.

If you continue the default/a specified suspension event has occurred [*delete as appropriate*] for 14 [*substitute* "7" *when using MW 98*] days after receipt of this notice, we may forthwith determine our employment under this contract without further notice.

Yours faithfully

函件 143
致雇主,在发出违约通知后终止雇佣关系
本函不适用于 GC/Works/1(1998)合同
专递/挂号邮件

Letter 143
To employer, determining employment after default notice
This letter is not suitable for use with GC/Works/1 (1998)
Special/recorded delivery

尊敬的先生:

就我方于[填入日期]向你方发出的违约通知,根据合同条件第 28.2.3 条[使用 IFC98 合同时,为"第 7.9.3 条";使用 MW98 合同时,为"第 7.3.1 条"]规定,请将本函视为通知,我方依据本合同并在不影响我方拥有的任何其他权利或索赔要求的情况下终止双方的雇佣关系。

我方正在安排从工地拆除临时建筑,运走施工设备以及材料。下周我方将就财务事宜再次致函你方。

你忠诚的

Dear Sir

We refer to the default notice sent to you on the [*insert date*].

Take this as notice that, in accordance with clause 28.2.3 [*substitute* "7.9.3" *when using IFC 98 or* "7.3.1" *when using MW 98*] we hereby determine our employment under this contract without prejudice to any other rights or remedies which we may possess.

We are making arrangements to remove all our temporary buildings, plant, etc. and materials from the site and we will write to you again within the next week regarding financial matters.

Yours faithfully

函件 144

致雇主,因雇主破产而终止雇佣关系
本函不适用于 GC/Works/1(1998)合同
专递/挂号邮件

Letter 144

To employer, determining employment on the employer's insolvency
This letter is not suitable for use with GC/Works/1 (1998)
Special/recorded delivery

尊敬的先生:

根据合同条件第28.3.3条[使用IFC98合同时,为"第7.10.3条";使用MW98合同时,为"第7.3.2条"]规定,由于[填入破产事件的准确细节],请将本函视为合同终止通知。你方收到本函后,终止即刻生效。

我方正在安排从工地拆除临时建筑,运走施工设备以及材料。下周我方将就财务事宜再次致函你方。本通知应不影响我方拥有的任何其他权利或索赔要求。

你忠诚的

Dear Sir

In accordance with the provisions of clause 28.3.3 [*substitute "7.10.3" when using IFC 98 or "7.3.2" when using MW 98*] of the conditions of contract, take this as notice that we hereby determine our employment, because [*insert precise details of the insolvency event*]. This determination will take effect on receipt of this notice.

We are making arrangements to remove all our temporary buildings, plant, etc. and materials from the site and we will write to you again within the next week regarding financial matters. This notice is without prejudice to any other rights and remedies we may possess.

Yours faithfully

函件 145
当任何一方提出合同终止时,致雇主
本函仅适用于 JCT98 合同或 IFC98 合同
专递/挂号邮件

Letter 145
To employer, where either party may determine
This letter is only suitable for use with JCT 98 or IFC 98
Special/recorded delivery

尊敬的先生:

由于发生以下事件[描述事件细节],全部未完工程或基本上全部未完工程的施工已连续中止一段时间[在附件中填入相应的时间段]。

根据合同条件第 28A.1 条[使用 IFC98 合同时,为"第 7.13.1 条"]规定,请将本函视为通知。根据本合同,我方与你方的雇佣关系将在你方收到本函后的 7 天内终止,除非该中止事件已结束。

终止之后的程序将根据合同条件第 28A.2 条~第 28A.7 条[使用 IFC98 合同时,为"第 7.14 条~第 7.19 条"]的规定进行。

你忠诚的

Dear Sir

The carrying out of the whole or substantially the whole of the uncompleted works has been suspended for a continuous period of [*insert the relevant period from the appendix*] by reason of the following event(s): [*insert details of the event(s)*].

Take this as notice, in accordance with the provisions of clause 28A.1 [*substitute* "7.13.1" *when using IFC 98*] of the conditions of contract, that our employment under the contract will determine within 7 days of the date of receipt of this notice unless, by that date, the suspension has been terminated.

Clauses 28A.2 to 28A.7 [*substitute* "7.14 to 7.19" *when using IFC 98*] will govern the procedure after that.

Yours faithfully

函件 146

致雇主,已投保的风险带来损害后决定终止雇佣关系
本函不适用于 MW98 或 GC/Works/1(1998)合同
专递/挂号邮件

Letter 146

To employer, determining employment after damage by insured risk
This letter is not suitable for use with MW 98 or GC/Works/1 (1998)
Special/recorded delivery

尊敬的先生:

  关于我方依据合同条件第22C.4条[使用IFC98合同时,为"第6.3C.4.1条"]规定,于[填入日期]发给你方的通知,我方认为,依据合同条件第22C.4.3.1条[使用IFC98合同时,为"第6.3C.4.4条"]的规定,终止我们的雇佣关系是公正合理的,因而,我方做出了立即终止雇佣关系的选择[自损失或损害发生之日起不少于28天内雇主必须收到终止通知函]。

你忠诚的

Dear Sir

We refer to our notice sent to you under the provisions of clause 22C.4 [*substitute "6.3C.4.1" when using IFC 98*] on [*insert date*]. We consider that it is just and equitable to determine our employment in accordance with clause 22C.4.3.1 [*substitute "6.3C.4.4" when using IFC 98*] and we hereby exercise our option to so determine forthwith. [*The determination must be received by the employer not less than 28 days from the date of the occurrence of the loss or damage*]

Yours faithfully

函件 147
致雇主,发出有意将争端事项诉诸裁决的通知
专递/挂号邮件

Letter 147
To employer, giving notice of intention to refer a dispute to adjudication
Special delivery

尊敬的先生:

根据合同条件第 41A.4.1 条 [使用 WCD98 合同时,为"第 39A.4.1 条";使用 IFC98 合同时,为"第 9A.4.1 条";使用 MW98 合同时,为"第 D4.1 条"或使用 GC/Works/1(1998)合同时,为"第 59(1)条"] 规定,我方打算将以下争端或分歧诉诸裁决:[填入争端内容]。我方将邀请裁决人[填入寻求补偿的内容,例如,"要求立即支付应付款 X 或支付裁决人裁定的应付款项"]。

[使用 GC/WORKS/1(1998)合同时,可添加以下内容:]

裁决人[填入姓名]为专用条款中规定的人员。

你忠诚的

抄送:指定机构[使用 GC/Works/1(1998)合同除外,因裁决人会获得一份副本]

Dear Sir

Under the provisions of clause 41A.4.1 [*substitute* "*39A.4.1*" *when using WCD 98,* "*9A.4.1*" *when using IFC 98,* "*D4.1*" *when using MW 98 or* "*59(1)*" *when using GC/Works/1 (1998)*] we intend to refer the following dispute or difference to adjudication: [*insert a description of the dispute*]. We will be requesting the adjudicator to [*insert the nature of the redress sought e.g,* "*order immediate payment of the outstanding amount of X or such sum as the adjudicator decides is due*"].

[*When using GC/Works/1 (1998), add*:]

The adjudicator will be [*insert name*] as specified in the abstract of particulars.

Yours faithfully
Copy: nominating body [*escept in the case of GC/Works/1 (1998) where the adjudicator should be sent a copy*]

函件 148

致指定机构,请求任命裁决人
本函不适用于 GC/Works/1(1998)合同
专递邮件

Letter 148

To nominating body, requesting nomination of an adjudicator
This letter is not suitable for use with GC/Works/1 (1998)
Special delivery

尊敬的先生:

　　现附上一份将本合同下的争端或分歧诉诸裁决的意向通知书。我方今天已将本通知书及随函送达合同的另一方:[填入雇主的名字]。本合同是以1998 JCT建筑合同标准文本为依据[使用 WCD98 合同时,为"JCT 承包商设计建筑合同标准文本";使用 IFC98 合同时,为"JCT 中型建筑合同标准文本"或 使用 MW98 合同时,为"JCT 小型建筑工程协议书"],包括补充条款第1~4条。你方为选定的裁决人提名机构。因此,根据合同条件第41A.2条[使用 WCD98 合同时,为"第39A.2条";使用 IFC98 合同时,为"第9A.2条"或使用 MW98 合同时,为"第D2条"]规定,我方向贵机构提出任命裁决人的申请。随函还附有已填好的申请表副本一份以及索赔人提交的金额为[填入金额]的支票一张。

你忠诚的

抄送:雇主
Dear Sir

We enclose a notice of intention to refer disputes and/or differences under the contract to adjudication. We have today served this notice and covering letter on the other party to the contract: [*insert the name of the employer*]. The contract was executed on JCT Standard Form of Building Contract 1998 [*substitute "JCT Standard Form of Building Contract With Contractor's Design" when using WCD 98, "JCT Intermediate Form of Building Contract" when using IFC 98 or "JCT Agreement for Minor Building Works" when using MW 98*] with Amendments 1 to 4 inclusive. You are the selected nominator. Therefore, in accordance with clause 41A.2 [*substitute "39A.2" when using WCD 98, "9A.2" when using IFC 98 or "D2" when using MW 98*], we hereby make application to you to appoint an adjudicator. A copy of the completed application form and the claimant's cheque in the sum of [*insert the amount*] is enclosed.

Yours faithfully
Copy: Employer

函件 149
致裁决人,附上提交裁决的争端事项
专递邮件

Letter 149
To adjudicator, enclosing the referral
Special delivery

尊敬的先生:

我方注意到,你是/已被任命为[视情形取舍]裁决人。现根据合同条件第41A.4.1条[使用 WCD98 合同时,为"第39A.4.1条";使用 IFC98 合同时,为"第9A.4.1条";使用 MW98 合同时,为"第D4.1条"或使用 GC/WORKS/1(1998)合同时,为"第59(2)条"]的规定,随函附上我方提交裁决的争端事项。提交裁决的争端事项包括争端或分歧的细节、我方观点的概述、我方寻求解脱或改正的说明以及希望裁决人给予考虑的其他材料。

提交裁决的争端事项的副本及有关文件已送交[填入雇主姓名,以及在使用 GC/Works/1(1998)合同时,加上项目经理和工料测量师的姓名]。

你忠诚的

抄送:雇主[使用 GC/Works/1(1998)合同时,加上项目经理和工料测量师]并附上附件

Dear Sir

We note that you are/have been appointed as [*delete as appropriate*] the adjudicator. In accordance with clause 41A.4.1 [*substitute* "39A.4.1" *when using WCD 98,* "9A.4.1" *when using IFC 98,* "D4.1" *when using MW 98 or* "59(2)" *when using GC/Works/1 (1998)*] we enclose our referral with this letter. Included are particulars of the dispute or difference, a summary of the contentions on which we rely, a statement of the relief or remedy sought and further material which we wish you to consider.

A copy of the referral and the accompanying documentation has been sent to [*insert name of employer and when using GC/Works/1 (1998) add the names of the PM and the QS*].

Yours faithfully
Copy: Employer [*add PM and QS when using GC/Works/1 (1998)*] with enclosures

函件 150
如果裁决人的裁决决定有利于我方,致雇主
专递邮件

Letter 150
To employer, if the adjudicator's decision is in your favour
Special delivery

尊敬的先生:

我方今天已收到裁决人裁决决定的副本。我方注意到,他裁决我方胜诉。因此,希望你方依据裁决人制定的时间表履行裁决决定[如果裁决人没有规定履行裁决的时间表,可填入"立即"一词以取代上面的画线部分]。

[视情形,可添加以下内容:]

因此,请[填入裁决人的决定以及有具体日期的时间表,例如"在2002年12月10日下班时支付总额为40000.00英镑的款项"]。

如果你方未能履行裁决人的决定,我方将立即向法庭提起诉讼。

你忠诚的

Dear Sir

We have today received a copy of the adjudicator's decision. We note that he has decided in our favour. You will be aware that you must comply with the adjudicator's decision in accordance with the timescale he has laid down [*if the adjudicator has not stated a time for compliance insert "promptly" in place of the words underlined*].

[*If appropriate, add*:]

Therefore, please [*insert the adjudicator's decisions and relevant timescales converted to actual dates, e.g, "pay us the sum of £40,000.00 by close of business on the 10 December 2002"*].

If you fail to comply with the adjudicator's decision, we will immediately take enforcement proceedings through the courts.

Yours faithfully

函件 151
致雇主，请求共同任命仲裁人
专递/挂号邮件

Letter 151
To employer, requesting concurrence in the appointment of an arbitrator
Special delivery

尊敬的先生：

现告知你方，依据我们双方于 [填入日期] 签订的合同条件第 7A 条 和第 41B 条 [使用 WCD98 合同时，为"第 6A 条和第 39B 款"；使用 IFC98 合同时，为"第 9A 条和第 9B 款"；使用 MW98 合同时，为"第 4 条和第 9 款"或使用 GC/Works/1(1998) 合同时，为"第 60 款"] 的规定，要求下面提到的双方之间的争端或分歧诉诸仲裁。请将本函视为要求共同任命仲裁人的请求。

提交仲裁处理的争端或分歧为：[简要陈述]

我方建议的下列三位人选供你方考虑，并且要求你方在本函送达之日起 14 天内同时完成任命工作。如若上述三人不能共同任命，我方将向英国皇家建筑师学会/皇家 Ulster 建筑师学会/皇家特许工料测量师学会/皇家仲裁人特许学会/法律学会/苏格兰法律学会/土木工程师学会[视情形取舍]的主席或副主席提请任命。

[列出三位人选的姓名和地址]

你忠诚的

Dear Sir

We hereby give you notice that we require the undermentioned dispute or difference between us to be referred to arbitration in accordance with article 7A and clause 41B [*substitute " article 6A and clause 39B" when using WCD 98, " article 9A and clause 9B" when using IFC 98, " article 4 and clause 9" when using MW 98 or " clause 60" when using GC/Works/1 (1998)*] of the contract between us dated [*insert date*]. Please treat this as a request to concur in the appointment of an arbitrator.

The dispute or difference is [*insert brief description*]

I propose the following three persons for your consideration and require your concurrence in the appointment within 14 days of the date of service of this letter, failing which we shall apply to the President or Vice-President of the Royal Institute of British Architects/Royal Society of Ulster Architects/Royal Institution of Chartered Surveyors/Chartered Institute of Arbitrators/Law Society/Law Society of Scotland/Institution of Civil Engineers [*delete as appropriate*].

[*List names and addresses of the three persons*]

Yours faithfully

函件 152
如果达不成共同任命仲裁人,致任命机构

Letter 152
To appointing body, if there is no concurrence in the appointment of an arbitrator

尊敬的先生:

我们是采用 JCT 98 合同文本的建筑承包商。该合同条件第 41B 条 [使用 WCD98 合同时,为"第 39.1 条";使用 IFC98 合同时,为"第 9B 条";使用 MW98 合同时,为"第 9.2 条"或使用 GC/Works/1(1998)合同时,为"第 60(1)条"]规定,若合同双方难以达成一致,请你方主席或副主席任命仲裁人。

烦请寄来相应的申请表和有关文件,同时告知申请所需的各项费用,不胜感激。

你忠诚的

Dear Sir

We are contractors who have entered into a building contract in JCT 98 form, clause 41B [*substitute* " *WCD 98 form, clause 39.1*", " *IFC 98 form, clause 9B*", " *MW 98 form, clause 9.2*" or " *GC/Works/1 (1998) form, clause 60(1)*" *as appropriate*] of which makes provision for your President or Vice-President to appoint an arbitrator in default of agreement.

We should be pleased to receive the appropriate form of application and supporting documentation, together with a note of the current fees payable on application.

Yours faithfully

函件 153

当工程实际竣工在即,致建筑师

Letter 153

To architect, if practical completion imminent

尊敬的先生:

  我方预计,工程将于[填入日期]竣工,请告知你方何时进行现场验收。届时我方将安排[填入姓名]先生到场,对任何询问给予当场解答。我方期待在检查后收到你方签发的实际竣工证书[使用WCD98合同时,代之以"工程已经实际完成的书面证明",或使用GC/Works/1(1998)合同时,代之以"按合同,工程已竣工的证书"]。

你忠诚的

Dear Sir

We anticipate that the Works will be complete on [*insert date*]. If you will let us know when you wish to carry out your inspection, we will arrange for Mr [*insert name*] to be on site to give immediate attention to any queries which may arise. We look forward to receiving your certificate of practical completion [*substitute* "*written statement that the Works have reached practical completion*" *when using WCD 98 or* "*certificate that the Works are completed in accordance with the contract*" *when using GC / Works / 1 (1998)*] following your inspection.

Yours faithfully

函件 154a
当竣工证书无理由被扣压时,致建筑师
本函不适用于 WCD98 合同

Letter 154a
To architect, if completion certificate wrongly withheld
This letter is not suitable for use with WCD 98

尊敬的先生:

你方[填入日期]的来函收悉。

得知你方认为工程[使用 GC/Works/1(1998)合同时,添加"依据合同的"工程]还没有达到实际竣工[使用 GC/Works/1(1998)合同时,代之以"竣工"],我方感到惊诧。你方所列举的未完成的项目应被认为是微不足道的。我方不认为这些项目会成为你方扣发竣工证书的理由。

我方强烈建议你方应重新考虑,并依照合同条件第 17.1 条[使用 IFC98 合同时,为"第 2.9 条";使用 MW98 合同时,为"第 2.4 条"或使用 GC/Works/1(1998)合同时,为"第 39 条"]的规定,立刻签发竣工证书。如果截止[填入日期],我方仍未收到你方的证书,即[填入日期]为实际竣工的日期[使用 GC/Works/1(1998)合同时,为"按合同工程已竣工之时"],我方将采取一切相应的措施,维护我方的权益。

我方将按正常方式处理你方列出的未完成的项目。

你忠诚的

Dear Sir

Thank you for your letter of the [insert date].

We are surprised to learn that, in your opinion, practical completion [substitute "completion" when using GC/Works/1 (1998)] of the Works [insert " in accordance with the contract" when using GC/Works/1 (1998)] has not been achieved. The items you list as outstanding can only be described as trivial. We do not consider that such items can possibly justify withholding your certificate.

We strongly urge you to reconsider the matter and to issue your certificate forthwith as required by clause 17.1 [substitute "2.9" when using IFC 98, "2.4" when using MW 98 or "39" when using GC/Works/1 (1998)] of the conditions of contract. If we do not receive your certificate by [insert date] naming [insert date] as the date of practical completion [substitute "when the Works were completed in accordance with the contract" when using GC/Works/1 (1998)], we will take whatever steps we deem appropriate to protect our interests.

The items you list are receiving attention in the normal way.

Yours faithfully

函件 154b

当竣工证书无理由被扣压时,致建筑师

本函仅适用于 WCD98 合同

Letter 154b

To architect, if completion statement wrongly withheld

This letter is only suitable for use with WCD 98

尊敬的先生:

你方[填入日期]的来函收悉。

得知你方认为工程尚未达到实际竣工,我方感到惊诧。你方所列举的未完成的项目应被认为是微不足道的,因而不能成为你方不发竣工证书的理由。

我方敦请你方注意这一事实,根据本合同,你方不具有签发证书的功能,而签发实际竣工的书面证明仅是记录实际情况的过程,并非要求你方的意见。我方强烈要求你方依据合同条件第 16.1 条的规定,立即签发书面说明。如果截止[填入日期]我方仍未收到你方的[填入证书名称]书面说明,即[填入日期]为实际竣工日期,我方将采取一切相应的措施,维护我方的权益。

我方将按正常方式处理你方列出的未完项目。

你忠诚的

Dear Sir

Thank you for your letter of the [*insert date*].

We are surprised to learn that, in your opinion, practical completion of the Works has not been achieved. The items you list as outstanding can only be described as trivial and they cannot possibly justify your contention.

We draw your attention to the fact that, under this form of contract, you have no certifying function and the issue of the written statement of practical completion is simply a process of recording a matter of fact. It is not something for your opinion. We strongly urge you to issue the written statement forthwith as required by clause 16.1 of the conditions of contract. If we do not receive your statement by [*insert date*] naming [*insert name*] as the date of practical completion, we will take whatever steps we deem appropriate to protect our interests.

The items you list are receiving attention in the normal way.

Yours faithfully

函件 155a
致雇主,同意占有部分已竣工工程
本函仅适用于 JCT98 合同或 WCD98 合同

Letter 155a
To employer, consenting to partial possession
This letter is only suitable for use with JCT 98 or WCD 98

尊敬的先生:

你方[填入日期]的来函收悉。我方同意你方占有部分已竣工工程[描述部分或若干部分已竣工工程]的要求。

1. 占有的日期为[填入日期],建筑师将代表你方依据合同条件第 18.1 条[使用 WCD98 合同时,将画线部分删除]规定签发书面证明。

2. [填入与当时移交情况相关的任何具体条件]。

如果你方或你方授权建筑师致函我方同意上述条件,我方将于[填入日期]安排移交相关的钥匙。[使用 WCD98 合同时,添加以下内容:]届时依据合同条件第 17.1 条的规定,我方将出具书面说明。

你忠诚的

抄送:建筑师

Dear Sir

In response to your letter of the [*insert date*], we consent to your request to take partial possession of [*describe part or parts*] provided:

1. The date for possession will be [*insert date*] and the architect will give a written statement to that effect on your behalf in accordance with clause 18.1 [*delete the underlined portion when using WCD 98*].

2. [*Insert whatever particular conditions may be appropriate to the circumstances*].
If you will, or you will authorise the architect, to write to us indicating agreement to the above conditions, we will make arrangements to hand over the appropriate keys on the [*insert date*]. [*If using WCD 98, add*:] We shall then issue a written statement in accordance with clause 17.1.

Yours faithfully
Copy: Architect

函件 155b
致雇主,同意占有部分已竣工工程
本函仅适用于 IFC98 合同

Letter 155b
To employer, consenting to partial possession
This letter is only suitable for use with IFC 98

尊敬的先生:

   你方[填入日期]的来函收悉。我方同意你方占有部分已竣工工程[描述部分或若干部分已竣工工程]的要求。

   1. 占有的日期为[填入日期],建筑师将代表你方依据合同条件第2.11条的规定,签发书面证明。

   2. [填入与当时移交情况相关的任何具体条件]。

   如果你方要求建筑师致函我方同意上述条件,我方将于[填入日期]做出必要安排移交相关的钥匙。

你忠诚的

抄送:建筑师

Dear Sir

In response to your letter of the [*insert date*], we consent to your request to take partial possession of [*describe part or parts*] provided:

1. The date for possession will be [*insert date*] and the architect will give a written statement to that effect on your behalf in accordance with clause 2.11.

2. [*Insert whatever particular conditions may be appropriate in the circumstances*].

If you will ask the architect to write to us agreeing to the above conditions, we will make the necessary arrangements to hand over the appropriate keys on the [*insert date*].

Yours faithfully

Copy: Architect

函件 156
致雇主,要求签发占有部分已竣工工程的书面证明
本函仅适用于 WCD98 合同

Letter 156
To employer, issuing written statement of partial possession
This letter is only suitable for use with WCD 98

尊敬的先生:

  同意我方[填入日期]函中阐明的条件的[填入日期]来函收悉。依据合同条件第 17.1 条的规定,本函为我方出具的书面证明。我方确认你方占有的部分或若干部分已竣工工程(相关部分)为[详细说明这些工程,避免出错]。你方占有的日期(相关日期)为[填入日期]。

  我方提请你方注意占有部分已竣工工程带来的后果,特别是涉及到缺陷责任期限的起始、保险责任以及任何我方可能承担的工程延误违约损害赔偿金的相应减少的情况。

你忠诚的

Dear Sir

Further to your letter of the [*insert date*] indicating agreement to the conditions contained in our letter of the [*insert date*], take this as the written statement which we are to issue in accordance with clause 17.1 of the conditions of contract. We identify the part(s) of the Works taken into possession (the relevant part(s)) as [*describe the part or parts in sufficient detail to allow no mistake*]. The date on which you took possession (the relevant date) was [*insert date*].

We draw your attention to the consequences of partial possession, particularly as they apply to the commencement of the defects liability period, your insurance liability and the reduction in any liability which we may have for liquidated damages.

Yours faithfully

函件 157
致雇主,不同意占有部分已竣工工程
本函不适用于 MW98 或 GC/Works/1(1998)合同

Letter 157
To employer, refusing consent to partial possession
This letter is not suitable for use with MW 98 or GC/Works/1 (1998)

尊敬的先生:

你方[填入日期]关于要求同意你方占有部分已竣工工程[说明部分或若干部分竣工工程]的函收悉。

遗憾地告知你方,我方认为不能同意你方这一要求,原因是[填入原因]。

[视情形,可添入以下内容:]

如果情况出现实质性变化,在我方能够同意你方此要求时,我方将会立即告知你方。

你忠诚的

抄送:建筑师

Dear Sir

Thank you for your letter of the [*insert date*] requesting our consent to you taking partial possession of [*describe part or parts*].

We regret that we feel unable to give our consent in this instance, because [*insert reasons*].

[*If appropriate, add*:]

We will let you know immediately if circumstances change so substantially that we feel able to consent to your request.

Yours faithfully

Copy: Architect

函件 158
收到缺陷清单后,致建筑师

Letter 158
To architect, after receipt of schedule of defects

尊敬的先生:

　　你方于[填入日期]发来的编号为[填入编号]的指令收悉。指令中列出了要求整改的工程缺陷清单,但是缺陷责任期 [使用 GC/Works/1(1998) 合同时,代之以"维修期"] 已结束。

　　我方已进行了初步检查,正在安排缺陷清单上的大多数项目的整改。然而,我方认为,下列项目不是我方的责任,原因如下:

[列出项目并给出原因]

　　当然,如果你方以书面形式确认可以计日工单价支付这些整改工程,我方将乐意完成这些整改工作。

你忠诚的

Dear Sir

Thank you for your instruction number [*insert number*] dated [*insert date*] scheduling the defects you require to be made good now that the defects liability period [*substitute "maintenance period" when using GC/Works/1 (1998)*] has ended.

We have carried out a preliminary inspection and we are making arrangements to make good most of the items on your schedule. However, we do not consider that the following items are our responsibility for the reasons stated:

[*List, giving reasons*]

We shall, of course, be happy to attend to such items if you will let us have your written agreement to pay us daywork rates for the work.

Yours faithfully

函件 159
当缺陷整改工作完成后,致建筑师

Letter 159
To architect, when making good of defects completed

尊敬的先生:

  很高兴地通知你方,根据你方的缺陷清单[视情形,代之以"你方的整改清单"]中所列的缺陷项目已全部整改完成,请你方亲自进行检查,[使用 JCT98 合同、IFC98 合同或 MW98 合同时,添加以下内容:]并签发缺陷整改完成证书以确认你方满意。

你忠诚的

Dear Sir

We are pleased to inform you that all making good of defects has been completed in accordance with your schedule [*if appropriate, substitute "your amended schedule"*]. We should be pleased if you would carry out your own inspection and confirm your satisfaction [*when using JCT 98, IFC 98 or MW 98, add*:] by issuing a certificate of completion of making good defects.

Yours faithfully

函件 160
致建筑师,在最后付款后返还图纸等文件材料
本函不适用于 WCD98 合同

Letter 160
To architect, returning drawings, etc. after final payment
This letter is not suitable for use with WCD 98

尊敬的先生:

　　你方[填入日期]的来函收悉。应你方要求,现附上我方拥有的标有你方姓名的图纸、详图、说明表以及其他类似的文件资料。当然,我方保留了合同文件副本,以备存档。

你忠诚的

Dear Sir

Thank you for your letter of [*insert date*].

We enclose, as requested, all copies of drawings, details, descriptive schedules and other documents of like nature which bear your name and which are in our possession. We have, naturally, retained our copy of the contract documents for record purposes.

Yours faithfully

第九章 分包商与分包合同

最后一章是处理与分包商的关系,涉及以下事宜:

- 权益转让
- 同意分包
- 意向函
- 工程担保
- 反对分包
- 不能签订分包合同
- 订立分包合同
- 分包合同保险事宜
- 分包工程图纸
- 指令(指示)
- 工期顺延
- 损失和(或)费用
- 分包工程款支付
- 拒付工程款的通知
- 暂停履约与合同终止
- 实际竣工
- 分包商的设计缺陷
- 分包合同的职业责任保险、担保和设计事宜
- 裁决

针对 JCT98 合同条件下的指定分包商和采用 IFC98 合同条件下的指定人员是非常复杂的,后面的许多信函提供了分别针对使用上述合同和使用 NSC/C 分包合同、NAM/SC 分包合同的示范格式。为了更好地使用这些标准格式的信函,必须仔细解读相应的分包合同条件及其他相关文件。

函件 161
致雇主,请求同意权益转让

Letter 161
To employer, requesting consent to assignment

尊敬的先生:

　　根据合同条款第 19.1 条[使用 WCD98 合同为"第 18.1.1 条",使用 IFC98 合同为和 mw98 合同"第 3.1 条"或使用 GC/work/1(1998)合同为"第 61 条"],我方十分高兴地获悉,你方已同意按本合同将我方应收的工程款转汇给[填入名称]。
　　由于[填入简单理由],我方才提出此请求。我方知道,本合同下我方的义务不会受到影响。

你忠诚的

Dear Sir

In accordance with clause 19.1 [*substitute* "18.1.1" *when using WCD 98*, "3.1" *when using IFC 98 and MW 98 or* "61" *when using GC/Works/1 (1998)*] we should be pleased to receive your consent to the assignment of our rights to payment under this contract to [*insert name*].

We wish to take such action, because [*give reasons briefly*]. We understand that our obligations under the contract will be unaffected.

Yours faithfully

函件 162
当请求雇主同意权益转让时,致雇主
本函件不适用 GC/Work/1(1998)合同

Letter 162
To employer, if asked to consent to assignment
This letter is not suitable for use with GC/Works/1 (1998)

尊敬的先生:

  谢谢你方[填入日期]的来函,从中我方获知你方希望根据本合同将你方的权益/部分权益[视情况取舍]转让。

  在可能情况下,我方十分想帮助你方。我方可能同意你方请求,这取决于某些保护措施。很明显,这并非我方能马上作出决定的事情,我方正在向专家咨询,希望不久再复函你方。

你忠诚的

Dear Sir

Thank you for your letter of the [*insert date*] from which we understand that you wish to assign your rights/certain of your rights [*delete as appropriate*] under the contract.
We are anxious to assist you if we can and we may be able to give our consent, subject to certain safeguards. Clearly, this is not something on which we can make an immediate decision and we are taking advice. We expect to be in a position to write to you again very soon.

Yours faithfully

函件 163

致建筑师,请求同意分包

Letter 163

To architect, requesting consent to sub-letting

尊敬的先生:

　　由于[说明理由],我方想将下述部分工程内容[使用 WCD98 合同时,若合适可填入"设计"工作]。希望你方能根据合同条款第 19.2.2 条[使用 WCD98 合同时,为"第 18.2.1/18.2.3 条";使用 IFC98 合同和 MW98 合同时,为"第 3.2 条"或使用 GC/Work/1(1998)时,为"第 62(1)条"]给予同意。

　　[填入将分包的部分工程内容或设计工作内容及分包商姓名]

你忠诚的

Dear Sir

We propose to sub-let portions of the work [*substitute* " *design*" *if appropriate when using WCD 98*] as indicated below, because[*state reasons*]. We should be pleased to receive your consent in accordance with clause 19.2.2 [*substitute* " *18.2.1/ 18.2.3*" *as appropriate when using WCD 98*, " *3.2*" *when using IFC 98 and MW 98 or* " *62(1)*" *when using GC/Works/ 1 (1998)*].

[*List the portions of the works or design and the names of the sub-contractors*]

Yours faithfully

函件 164
致雇主,请求按第 19.3.1 条,在人员名单中增加人员
本函件仅适用于使用 JCT98 合同

Letter 164
To employer, requesting consent to addition of persons to clause 19.3.1 list
This letter is only suitable for use with JCT 98

尊敬的先生:

　　根据合同条款第 19.3.2.1 条,我方切盼你方能同意将[填入姓名]列入到从事[填入工作内容]工作的合同支付帐单[填入帐单页码及相关内容]中的人员名单之中。

你忠诚的

抄送:建筑师

Dear Sir

In accordance with clause 19.3.2.1 of the conditions of contract, we should be pleased to receive your consent to the addition of [*insert name*] to the list of persons named in the contract bills [*insert page number and reference*] for [*insert description of work*].

Yours faithfully

Copy: Architect

函件 165

致雇主,同意按第 19.3.1 条,在人员名单中增加人员
本函件仅适用于使用 JCT98 合同

Letter 165

To employer, giving consent to addition of person to clause 19.3.1 list
This letter is only suitable for use with JCT 98

尊敬的先生:

  谢谢你方[填入日期]的来函。
  我方同意将[填入姓名]列入到从事[填入工作内容]工作的合同支付帐单[填入帐单页码及相关内容]的人员名单之中。

你忠诚的

抄送:建筑师

Dear Sir

Thank you for your letter of the [*insert date*].

We consent to the addition of [*insert name*] to the list of persons named in the contract bills [*insert page number and reference*] for [*insert description of work*].

Yours faithfully

Copy: Architect

函件 166
致分包商：意向函
专递/挂号邮件

Letter 166
To sub-contractor: letter of intent
Special / recorded delivery

尊敬的先生：

你方[填入日期]就[填入工程性质或设计内容]的报价已接受,我方有意在主合同文件顺利签订后与你方签订分包合同。

我方并不认为,本意向函本身或本函连同你方报价应视为具有约束力的合同。但我方准备指令你方开始工作[填入有限的工作内容或设计工作]。若因某种原因,工程不再进行,我方只承诺支付你方已从事的工作[填入有限的工作内容或设计工作],支付的依据是上述你方报价单中的单价。

只有得到进一步的书面指令,你方才可开展你方报价中的其他工作内容[若合适,可填入"设计工作"]。无论何种情况下,我方不承担进一步的责任。

你忠诚的

Dear Sir

Your quotation of the [*insert date*] for [*insert nature of the work or design*] is acceptable and we intend to enter into a sub-contract with you after the main contract documents have been satisfactorily executed.

It is not our intention that this letter, taken alone or in conjunction with your quotation, should form a binding contract. However, we are prepared to instruct you to [*insert the limited nature of the work or design required*]. If, for any reason, the project does not proceed, our commitment will be strictly limited to payment for [*insert the limited nature of the work or design required*]. The basis of payment will be the prices in your quotation noted above.

No other work [*substitute "design" if appropriate*] included in your quotation must be carried out without a further written order. No further obligation is placed upon us under any circumstances.

Yours faithfully

函件 167

致国内分包商,当招标邀请信中已提及时,要求提供担保

Letter 167

To domestic sub-contractor, requiring a warranty if noted in invitation to tender

---

尊敬的先生:

  我方已收到你方[填入日期]就[填入工程内容或设计工作]的报价,报价总额为[填入数额]。我方接受你方报价并附上正式文件资料,请根据我方招标邀请信,完成你方报价的全部内容。

  我方同时附上担保格式,你方应把它作为契约签字/履行[视情况取舍]并与合同文件一并返给我方。该担保格式与招标邀请信中的保函格式相同。

  请在[填入日期]之前将填完的文件资料送我方,以便我方完成并将一份正本送你方。

你忠诚的

Dear Sir

We are in receipt of your quotation dated [insert date] in the sum of [insert amount] for [insert the nature of the work or design] which we have pleasure in accepting and we enclose the formal documentation for completion all in accordance with our invitation to tender and your quotation.

We also enclose the form of warranty which you should sign/execute as a deed [delete as appropriate] and return to us with the contract documents. This is exactly the same form of warranty which was attached to the invitation to tender.

Please let us have the completed documents by [insert date] when we will arrange completion and send a true copy to you.

Yours faithfully

函件 168
致国内分包商,当招标邀请信中未提及时,要求提供担保

Letter 168
To domestic sub-contractor, requiring a warranty if not noted in the invitation to tender

尊敬的先生:

我方谨提及你方[填入日期]提交的总价为[填入数额]的工程[填入工程内容或设计工作]报价。

若你方在所附的担保书上签字/履行担保书义务[视情况取舍]作为把雇主/将来的承租人/出资人[视情况取舍]作为受益人的一种契约,我方就准备与你方签订分包合同。这是雇主的要求,只有当担保生效时,雇主才同意合同分包。

我方期待着在[填入日期]之前收到你方填好的担保格式以便我方继续完成合同文件工作。

你忠诚的

Dear Sir

We refer to your quotation dated [*insert date*] in the sum of [*insert amount*] for [*insert the nature of the work or design*].

We should be prepared to enter into a sub-contract with you if you will sign the attached warranty/execute the attached warranty as a deed [*delete as appropriate*] in favour of the employer/future tenants/funders [*delete as appropriate*]. This is a requirement of the employer who will not give his consent to sub-letting unless a warranty is in place.

We look forward to receiving the completed warranty form by [*insert date*] so that we may proceed with the contract documentation.

Yours faithfully

函件 169
当国内分包商不愿提供担保时,致建筑师

Letter 169
To architect, if domestic sub-contractor refuses to provide a warranty

尊敬的先生:

　　[填入分包商姓名]不愿按照你方建议的条件提供担保。

　　[加入下一条]

　　很明显他不准备提供担保,如你方所知,我方的问题是有能力承担这项工程/设计[视情况取舍]的分包商数量十分有限,而工程又十分需要这家分包商。请给予指示,我方将十分感激。

　　[或加入]

　　很明显,若对担保条件作修改,他是准备提供担保的。现附上经修改的担保函的样本。我方将高兴地收到你方对修改的担保条件满意的确认。

你忠诚的

Dear Sir

[*Insert name of sub-contractor*] refuses to enter into a warranty on the terms which you have proposed.

[*Add either*:]

It appears that he is not prepared to enter into any warranty. Our problem is, as you are aware, that the number of sub-contractors who are able to carry out this kind of work/design [*delete as appropriate*] is extremely limited and this contractor is in great demand. We should be pleased to have your observations and instructions.

[*Or:*]

It appears that he is prepared to enter into a warranty if the terms are amended and we enclose an example of such a warranty. We should be glad to have your confirmation that the amended terms are satisfactory.

Yours faithfully

函件 170a

致建筑师,反对指定分包商
本函件仅适用于使用 JCT98 合同

Letter 170a

To architect, objecting to a nominated sub-contractor
This letter is only suitable for use with JCT 98

尊敬的先生:

收到你方根据合同条件第 35.6 条所发出的关于分包商的指定的指示。

根据第 35.5.1 条,我方有适当的理由不同意对[填入姓名]的指定。我方不同意的理由是[解释理由]。

请遵照第 35.5.2 条内容,采取下列任一种处理办法:

1. 签发进一步指令,排除我方的不同意;或
2. 取消以前的指令,即:
   a) 根据第 13.2 条,发出指令取消该工作,或
   b) 根据第 35.6 条,发出指令指定另一家分包商。

你忠诚的

Dear Sir

We are in receipt of your instruction on Nomination NSC/N nominating a sub-contractor in accordance with clause 35.6 of the conditions of contract.

We have reasonable objection, under clause 35.5.1, to the nomination of [*insert name*]. The reason for our objection is [*explain*].

Please operate the provisions of clause 35.5.2 and either:

1. Issue further instructions to remove our objection; or
2. Cancel the instruction and either:
   a) Issue an instruction under clause 13.2 to omit the work, or
   b) Issue an instruction under clause 35.6 to nominate another sub-contractor.

Yours faithfully

函件 170b

致建筑师,反对指定人员

本函件仅适用于使用 IFC98 合同

Letter 170b

To architect, objecting to a named person

This letter is only suitable for use with IFC 98

尊敬的先生:

  收到你方[填入日期]编号为[填入编号]的指令,要求我方与[填入姓名]签订分包合同。

  根据第 3.3.2(c)条,我方有适当的理由不同意签订此分包合同,我方反对的理由为[填入理由]。

你忠诚的

Dear Sir

We are in receipt of your instruction number [*insert number*] dated [*insert date*] instructing us to enter into a sub-contract with [*insert name*].

We have reasonable objection, under clause 3.3.2(c), to entering into such sub-contract. The reason for our objection is [*explain*].

Yours faithfully

函件 170c

致建筑师,反对指定分包商

本函件仅适用于使用 GC/Work/1(1998)合同

Letter 170c

To architect, objecting to a nominated sub-contractor

This letter is only suitable for use with GC/Works/1 (1998)

尊敬的先生:

　　收到你方[填入日期]编号为[填入编号]的指令,要求我方与[填入姓名]就[填入工程内容]工程签订分包合同。

　　由于[说明理由],我方有适当的理由不同意雇用该指定分包商。反对指定分包商的决定是根据合同条件第 63(6)条作出的。

你忠诚的

Dear Sir

We are in receipt of your instruction number [*insert number*] dated [*insert date*] instructing us to enter into a sub-contract with [*insert name*] for [*insert nature of work*] work.

We have reasonable objection to the employment of such nominated sub-contractor, because[*state reasons*]. This objection is made under the provisions of clause 63(6) of the conditions of contract.

Yours faithfully

函件 171a

当承包商不能与指定分包商达成协议或不能签订分包合同时，致建筑师

本函件仅适用于使用 JCT98 合同

Letter 171a

To architect, if contractor is unable to reach agreement or conclude a sub-contract with the nominated sub-contractor

This letter is only suitable for use with JCT 98

尊敬的先生：

　　就你方于[填入日期]根据合同条件第35.6条发出的关于指定分包商的指定的指示，本函就是根据第35.8条的书面通知。

　　[然后加入]

　　我方不能按 NSC/T 第三部分完成/履行 NSC/A 协议 [视情况取舍] 在规定的10天内与分包商达成协议，但我方能按照第35.8条规定在[填入日期]之前与分包商达成协议。

　　[或]

　　我方不能按 NSC/T 第三部分完成/履行 NSC/A 协议 [视情况取舍] 在规定的10天内与分包商达成协议。这是因为[填入理由]。

你忠诚的

Dear Sir

We refer to your instruction on Nomination NSC/N dated [*insert date*] under the provisions of clause 35.6 of the conditions of contract. This letter is notice in writing in accordance with clause 35.8

[*Then add either*:]

that we have been unable to complete NSC/T Part 3/execute agreement NSC/A [*delete as appropriate*] with the sub-contractor within the 10 days stipulated, but we expect to be able to comply with clause 35.8 by [*insert date*].

[*Or*:]

that we have been unable to complete NSC/T Part 3/execute agreement NSC/A [*delete as appropriate*] with the sub-contractor within the 10 days stipulated. This is due to [*insert reasons*].

Yours faithfully

函件 171b
当承包商不能与指定人员签订分包合同时,致建筑师
本函件仅适用于使用 WCD 98 合同

Letter 171b
To architect, if contractor unable to enter into a sub-contract with named person
This letter is only suitable for use with WCD 98

尊敬的先生:

　　根据补充条款第 S4.2.1 条,我方正努力与"雇主要求"[填入页数及条款号]中指定的[填入姓名]签订分包合同。但我方的努力付诸东流,因为[填入理由],因此,若你方能按第 S4.2.2 条执行,我方将十分高兴:
　　1. 签发变更指令,修改"雇主要求"中的该条内容以便我方能签订分包合同;或
　　2. 签发变更指令,取消该指定的分包工程,然后再签发指令要求实施该项工作。

你忠诚的

Dear Sir

In accordance with supplementary provision S4.2.1, we have attempted to enter into a sub-contract with [*insert name*] who was named in the Employer's Requirements at [*insert reference to page and item number*]. Our efforts have been unsuccessful, because [*insert reason*] and we should be pleased if you would operate the provisions of provision S4.2.2 and either:

1. Issue a change instruction to amend the item in the Employer's Requirements so that we can enter into the sub-contract; or

2. Issue a change instruction to omit the named sub-contract work and issue further instructions about the carrying out of that work.

Yours faithfully

函件 171c
当根据合同特殊条款不能与指定人员签订分包合同时,致建筑师
本函件仅适用于使用 IFC 98 合同

Letter 171c
To architect, if unable to enter into sub-contract with named person in accordance with particulars
This letter is only suitable for use with IFC 98

尊敬的先生:

  根据合同条件第 3.3.1 条,在此通知你方,根据合同文件的内容,我方不能与[填入姓名]签订分包合同。下列特殊条款内容表明不能签订这类分包合同的理由:

[说明特殊条款具体内容]

若你方能按照合同要求签发指令,我方将十分高兴。

你忠诚的

Dear Sir

We hereby notify you in accordance with clause 3.3.1 of the conditions of contract that we are unable to enter into a sub-contract with [*insert name*] in accordance with the particulars given in the contract documents. The following are the particulars which have prevented the execution of such sub-contract:

[*Specify particulars*]

We should be pleased if you would issue your instructions as required under the contract.

Yours faithfully

函件 172
当建筑师不能证实不一致时,致建筑师
本函件仅适用于使用 JCT 98 合同

Letter 172
To architect, who considers non-compliance unjustified
This letter is only suitable for use with JCT 98

尊敬的先生:

　　就我方根据合同条件第 35.8 条于[填入日期]发出的通知,我方确认已收到你方于[填入日期]的答复。我方注意到你方并不认为我方通知中所提及的事项已证实了与分包商指定的指令不一致。

　　因此,我方仅想说明的是,正因为那位指定的分包商不准备履行 NSC/A 协议,其原因在某种程度上已阐明。在此情形下,第 35.9.2 条中的第一段无法履行,因为我方无法给[填入姓名]施加压力,我方认为也无必要施加压力。我方建议你方说服[填入姓名]履行该协议或你方按第 35.9.2 条的第二段自行实施。

　　同时,我方必须指出,若此事在[填入日期]不能最后解决,我方有权要求补偿我方由此所蒙受的工期延误和经济损失或(和)产生的额外费用。

你忠诚的

Dear Sir

Following our notice dated [*insert date*] in accordance with clause 35.8 of the conditions of contract, we have received your response of the [*insert date*]. We note that you do not consider that the matters identified in our notice justify non-compliance with the nomination instruction.

As to that, we would merely state that it is the nominated sub-contractor who is not prepared to execute the agreement NSC/A for the reasons we have already set out at some length. In the circumstances, the first paragraph of clause 35.9.2 is inoperable, because we know of no pressure which we can bring to bear on [*insert name*] nor do we think it would be desirable. We suggest that you either persuade [*insert name*] to execute the agreement or you operate the second paragraph of clause 35.9.2.

Meanwhile, we must notify you that if the matter is not finalised by [*insert date*], we shall suffer delay and loss and/or expense for which we will be entitled to reimbursement.

Yours faithfully

函件 173
当指定的供应商不愿签订合适的采购合同时,致建筑师
本函件仅适用于使用 JCT 98 合同

Letter 173
To architect, if nominated supplier not willing to enter into a suitable contract of sale
This letter is only suitable for use with JCT 98

尊敬的先生:

  我方收到[填入日期]你方编号为[填入编号]的指令,要求我方为[填入货物名称]向[填入姓名]签订采购合同。

  我方于[填入日期]向该供应商发了函,但我方不得不告知你方,他们不愿依据包含第 36.4 条内容的条款签订供货合同。他们反对的具体条款为[填入条款编号]。

  若你方能书面确认将上述条款中的责任作出限定/限制/排除[视情况取舍]并附上雇主的书面确认书,即我方可能因此而蒙受的经济损失或(和)产生的额外费用可以得到补偿,我方则准备签订该供货合同,请参见合同条件第 36.5 条。

你忠诚的

Dear Sir

We are in receipt of your instruction number [*insert number*] dated [*insert date*] from which we note that you wish us to place an order with [*insert name*] for [*insert nature of goods*].

We wrote to this supplier on the [*insert date*] and we have to report that they are unwilling to enter into a contract of sale on terms which include the provisions of clause 36.4. The particular term to which they object is [*insert number*].

If you will specifically approve in writing the above restriction/limitation/exclusion [*delete as appropriate*] of liability together with the employer's written assurance that we will be reimbursed any loss and/or expense which we may incur as a result, we will be prepared to enter into such contract of sale. Clause 36.5 of the conditions of contract refers.

Yours faithfully

函件 174
当所推荐的分包商撤回其报价时,致建筑师
本函件仅适用于使用 JCT 98 合同

Letter 174
To architect, if proposed sub-contractor withdraws his offer
This letter is only suitable for use with JCT 98

尊敬的先生:

　　就收到你方[填入日期]根据合同条件第35.6条而签发的编号为[填入编号]的指令,我方确认在今日早些时候的电话交谈中已告知,我方刚收到所推荐的分包商[填入姓名]于[填入日期]的来函,要求撤回其报价,详见附信。

　　显而易见,除了请求你的指示以避免工程进度延误外,我方实在无能为力,你方或雇主可以根据 NSC/T 第二部分的规定和根据 NSC/W 第1.1条对分包商施加压力,尽管我方对此是否有效仍有疑虑。无论如何,一个勉强的分包商是没有什么实际作用的。

　　我方认为,这些情况并不十分符合第35.8条所预见的后果。但若你方不同意,请将本函视作根据该条款所发出的通知。

你忠诚的

Dear Sir

Further to receipt of your instruction number [*insert number*] dated [*insert date*] which you issued under the provisions of clause 35.6 of the conditions of contract, we confirm our telephone conversation earlier today that we have just received the enclosed letter dated [*insert date*] from the proposed sub-contractor, [*insert name*], withdrawing his offer.

It appears that there is little we can do except to request your instructions in order to avoid delay to the progress of the work. It may be that you or the employer can exert pressure on the sub-contractor by virtue of NSC/T Part 2, Stipulations and NSC/W clause 1.1, although it seems doubtful to us. In any event, a reluctant sub-contractor is of little practical use.

We do not consider that these circumstances quite fit into the sequence envisaged by clause 35.8, but if you disagree, please consider this as notice under the provisions of that clause.

Yours faithfully

函件 175
当承包商与指定分包商签订分包合同后,致建筑师
本函件仅适用于使用 JCT 98 合同

Letter 175
To architect, if contractor enters into a sub-contract with nominated sub-contractor
This letter is only suitable for use with JCT 98

尊敬的先生:

根据合同条款第 35.7 条,随函附上所签订的 NSC/A 协议以及所签订的 NSC/T 第三部分的副本。

你忠诚的

Dear Sir

In accordance with clause 35.7 of the conditions of contract, we enclose a copy of the completed agreement NSC/A and of the agreed and signed NSC/T Part 3.

Yours faithfully

函件 176
当承包商与指定的人员签订分包合同后,致建筑师
本函件仅适用于使用 WCD 98 合同

Letter 176
To architect, if contractor enters into a sub-contract with named person
This letter is only suitable for use with WCD 98

尊敬的先生:

  根据补充协议第 S4.2.1 条的规定,特函告你方,我方已于[填入日期]与[填入姓名]签订了分包合同。

你忠诚的

Dear Sir

This letter is to notify you, as required by supplementary provision S4.2.1, that we entered into a sub-contract with [*insert name*] on the [*insert date*].

Yours faithfully

函件 177
当承包商与指定的人员签订分包合同时,致建筑师
本函件仅适用于使用 IFC 98 合同

Letter 177
To architect, if contractor enters into contract with named person
This letter is only suitable for use with IFC 98

尊敬的先生:

根据合同条件第 3.3.1 条,我方必须通知你方,我方已于 [填入日期] 与 [填入姓名] 签订了分包合同。

你忠诚的

Dear Sir

In accordance with clause 3.3.1 of the conditions of contract, we must inform you that we entered into a sub-contract with [*insert name*] on the [*insert date*].

Yours faithfully

函件 178
关于保险事宜,致分包商

Letter 178
To sub-contractor, regarding insurance

尊敬的先生:

　　根据分包合同第 6.5 条[使用 NAM/SC 合同时,为"第 8 条";使用 DOM/Z 或 GC/Work/SC 合同时,为"第 7 条"]的有关保险的要求,请务必于[填入日期]之前提交保险单及保险费交付的收据。

你忠诚的

Dear Sir

Please submit insurance policies and premium receipts in respect of the insurance which you are required to maintain under clause 6.5 [*substitute " 8" when using NAM/SC, "7" when using DOM/2 or GC/Works/SC*] of the sub-contract. The policies and receipts must be in our hands by [*insert date*].

Yours faithfully

函件 179
当分包商不能办理保险时,致分包商

Letter 179
To sub-contractor, if he fails to maintain insurance cover

尊敬的先生:

就今天与你方 [填入姓名] 先生的电话交谈中获悉,你方不能提供按分包合同第 6.5 条[使用 NAM/SC 合同时,为第 8 条;使用 DOM/Z 或 GC/Work/SC 合同时,为第 7 条]要求的已办理并保持有效的保险文件证明。

鉴于保险的重要性以及根据第 1.10 条,第 6.2 条及第 6.3 条[使用 NAM/SC 合同时为"第 6 和 7 条";使用 DSC/C 合同时,为"第 1.11,6.2 和 6.3 条";或使用 DOM/2 或 GC/Work/SC 合同时为"第 5 和 6 条"]我方将根据第 6.8 条[使用 NAM/SC 合同时,为"第 10.2 条";或使用 DOM/2 或 GC/Work/SC 合同时为"第 9.2 条"]立即行使我方权力。我方所支付的任何保险费都将从你方的应付款或将付款中扣除,或作为你方欠款而收回。

你忠诚的

Dear Sir

We refer to our telephone conversation today with your Mr [*insert name*] and confirm that you are unable to produce documentary evidence that the insurance required by clause 6.5 [*substitute* "*8*" *when using NAM/ SC or* "*7*" *when using DOM/ 2 or GC/ Works/ SC*] of the sub-contract has been properly effected and maintained.

In view of the importance of the insurance and without prejudice to your liabilities under clauses 1.10, 6.2 and 6.3 [*substitute* "*6 and 7*" *when using NAM/ SC,* "*1.11, 6.2 and 6.3*" *when using DSC/ C or* "*5 and 6*" *when using DOM/ 2 or GC/ Works/ SC*], we are arranging to exercise our rights under clause 6.8 [*substitute* "*10.2*" *when using NAM/ SC or* "*9.2*" *when using DOM/ 2 or GC/ Works/ SC*] immediately. Any sum or sums payable by us in respect of premiums will be deducted from any money due or to become due to you or will be recovered from you as a debt.

Yours faithfully

函件 180

致分包商,并附上图纸

Letter 180

To sub-contractor, enclosing drawings

尊敬的先生:

[若使用 NAM/SC 合同,信函开头为:]

根据分包合同第 2.3 条:

[然后,否则信函开头为:]

现附上每份图纸 [填入图号] 的两份复印件,我方认为,这些图纸足以使你方实施并完成分包合同的工程。

你忠诚的

Dear Sir

[*If using NAM/SC, begin*:]

In accordance with clause 2.3 of the sub-contract,

[*Then, or otherwise begin*:]

We enclose two copies of each of drawings numbered [*insert numbers*] which we consider to be reasonably necessary to enable you to carry out and complete the sub-contract works.

Yours faithfully

函件 181a

当要求批准分包时,致指定分包商

本函件仅适用于使用 NSC/C 合同

Letter 181a

To nominated sub-contractor, if asked to consent to assignment

This letter is only suitable for use with NSC/C

尊敬的先生:

谢谢你方[填入日期]的来函,从中我方得知你方想与[填入姓名]签订分包合同。

根据分包合同第 3.13 条,我方已将你方来函的副本交建筑师征求意见。由于此事关系重大,我方不能立即作出决定,但我方将很快再函告你方。

你忠诚的

Dear Sir

Thank you for your letter of the [insert date] from which we understand that you wish to assign the sub-contract to [insert name].

We have sent a copy of your letter to the architect under the provision of clause 3.13 of the sub-contract and we are taking advice. This is a serious matter and we cannot make an immediate decision, but we expect to be in a position to write to you again very soon.

Yours faithfully

函件 181b

当要求批准分包时,致分包商

本函件仅适用于使用 NAM/SC、DSC/C、DOM/2 或 GC/Works/SC 合同

Letter 181b

To sub-contractor, if asked to consent to assignment

This letter is only suitable for use with NAM/SC, DSC/C, DOM/2 or GC/Works/SC

尊敬的先生:

  谢谢你方[填入日期]的来函,从中我方获悉你方想与[填入姓名]签订分包合同。
  此事关系重大,因此我方不能立即作出决定。我方正在征求意见并在不久即可函告你方。

你忠诚的

Dear Sir

Thank you for your letter of the [*insert date*] from which we understand that you wish to assign the sub-contract to [*insert name*].

This is a serious matter and we cannot make an immediate decision. We are taking advice and we expect to be in a position to write to you again very soon.

Yours faithfully

函件 182a
当分包未获批准时,致指定分包商
本函仅适用于使用 NSC/C 合同

Letter 182a
To nominated sub-contractor, if he sub-lets without consent
This letter is only suitable for use with NSC/C

尊敬的先生:

  我方得知你方已将[填入分包工程的内容]分包给[填入姓名]。
  由于建筑师及我方均未同意,你方的行为均成了对分包合同第 3.14 条的违约,因此必须停止。请回函确认你方将按此函要求执行,否则根据主合同第 35.24.1 条的要求,我方将通知建筑师根据分包合同第 7.1.4 条你方已成违约。

你忠诚的

Dear Sir

We are informed that you have sub-let [*insert the part of the works sub-let*] to [*insert name*]

Since neither we nor the architect have given our consent, your action is in breach of sub-contract clause 3.14 and must cease forthwith. Please confirm, by return, that you will comply with this letter otherwise we will inform the architect, as required by clause 35.24.1 of the main contract, that you have made default in respect of the matter referred to in clause 7.1.4 of the sub-contract.

Yours faithfully

函件 182b

当分包未获批准时,致分包商

本函件仅适用于使用 NAM/SC、DSC/C、DOM/2 或 GC/Works/SC 合同

Letter 182b

To sub-contractor, if he sub-lets without consent

This letter is only suitable for use with NAM/SC, DSC/C, DOM/2 or GC/Works/SC

尊敬的先生:

我方得知你方已将[填入分包工程内容]分包给[填入姓名]。

由于我方尚未同意,你方的行为构成了对分包合同第 24.2 条 [使用 DSC/C 或 DOM/2 合同时,为"第 3.13.1 条",或使用 GC/Work/SC 合同时,为"第 26.2 条"]的违约,因此必须停止。请回函确认你方已按此函要求执行,否则我方将根据第 27.2.1 条 [使用 DSC/C 合同时,为"第 7.1.1 条",使用 DOM/2 合同时为"第 29.2 条",或 GC/Work/SC 合同时,为"第 29.1.2 条"]终止对你方的雇用。

你忠诚的

Dear Sir

We are informed that you have sub-let [*insert the part of the works sub-let*] to [*insert name*].

Since we have not given our consent, your action is in breach of the sub-contract clause 24.2 [*substitute* "3.13.1" *when using DSC/C or DOM/2 or* "26.2" *when using GC/Works/SC*] and must cease forthwith. Please confirm, by return, that you will comply with this letter otherwise we will take action to terminate your employment under clause 27.2.1 [*substitute* "7.1.1" *when using DSC/C,* "29.2" *when using DOM/2 or* "29.1.2" *when using GC/Works/SC*].

Yours faithfully

函件 183a

致指定分包商,同意分包

本函件仅适用于使用 NSC/C 合同

Letter 183a

To nominated sub-contractor, giving consent to sub-letting

This letter is only suitable for use with NSC/C

尊敬的先生:

  我方收到你方[填入日期]的来函,要求批准你方将[填入部分工程名称]分包给[填入姓名]。

  我方与建筑师商量后,建筑师同意你方要求,因此,你方可以将该部分工程分包给上述人员/公司[视情况取舍]。

你忠诚的

Dear Sir

We are in receipt of your letter of the [*insert date*] in which you request consent to the sub-letting of [*insert the portion of works*] to [*insert name*].

We have consulted the architect, who consents to your request. You may, therefore, sub-let such portion of the sub-contract works to such person/firm [*delete as appropriate*] noted above.

Yours faithfully

函件 183b

致分包商,同意分包

本函件仅适用于使用 NAM/SC、DSC/C、DOM/2 或 GC/Works/SC 合同

Letter 183b

To sub-contractor, giving consent to sub-letting

This letter is only suitable for use with NAM/SC, DSC/C, DOM/2 or GC/Works/SC

尊敬的先生:

  为答复你方[填入日期]的来函,我方十分高兴告知你方,根据分包合同第24.2条[使用DSC/S合同时,为"第3.13.1条";使用DOM/2合同或GC/Works/SC合同时,为"第26.2条"],我方同意将[填入分包工程内容]分包给[填入姓名]。

你忠诚的

Dear Sir

In response to your letter of the [*insert date*], we are pleased to give our consent, under the provisions of clause 24.2 [*substitute* "3.13.1" *when using DSC/ or* "26.2" *when using DOM/2 or GC/Works/SC*] of the sub-contract, to the sub-letting of [*insert portion of the works*] to [*insert name*].

Yours faithfully

函件 184

致分包商,要求服从指示
本函件仅适用于使用 NSC/C、NAM/SC、DSC/C、DOM/2 或 GC/Works/SC 合同
专递/挂号邮件

Letter 184

To sub-contractor, requiring compliance with direction
This letter is only suitable for use with NSC/C, NAM/SC, DSC/C, DOM/2 or GC/Works/SC
Special/recorded delivery

尊敬的先生:

　　根据分包合同条件第 3.10 条 [使用 NAM/SC 分包合同时,为"第 5.4 条";使用 DOM/2 分包合同时,为第"4.5 条";使用 GC/Works/SC 分包合同时,为"第 4.1 条"],我方要求你方执行我方[填入日期]发出的[填入编号]指令。请将此函作为正式通知,现再次附上指令的副本。

　　若收到本通知 7 天内你方尚未执行,我方将[使用 NSC/C 合同时,填入"征求建筑师同意后"]雇用他人并支付费用来执行此指令,与此雇用所发生的有关费用将从你方根据分包合同到期应收款中扣除或作为你方欠款而收回。

你忠诚的

抄送:建筑师

Dear Sir

Take this as notice under clause 3.10 [*substitute* "5.4" *when using NAM/SC,* "4.5" *when using DOM/2 or* "4.1" *when using GC/Works/SC*] of the conditions of sub-contract that we require you to comply with our direction number [*insert number*] dated [*insert date*], a further copy of which is enclosed.

If within 7 days of receipt of this notice, you have not begun to comply we will [*when using NSC/C, insert* "seek the architect's permission to"] employ and pay others to comply with such direction. All costs incurred in connection with such employment will be deducted from money due or to become due to you under the sub-contract or will be recovered from you as a debt.

Yours faithfully

Copy: Architect

函件 185
当承包商不同意所谓的口头指令时,致分包商
本函件仅适用于使用 NSC/C、DSC/C 或 DOM/2 合同

Letter 185
To sub-contractor, contractor dissents from alleged oral direction
This letter is only suitable for use with NSC/C, DSC/C or DOM/2

尊敬的先生:

收到你方[填入日期]的来函,意在要求确认你方声称的由[填入姓名]在施工现场/打电话[视情况取舍]发出的口头指令。

根据第3.3.3条[使用DSC/C合同时,为"第3.3条";或使用DOM/Z合同时,为"第4.4条"],在此我方正式表示不同意此指令,同时申明所谓的口头指令未曾发出或根本未发出过。

你忠诚的

Dear Sir

We are in receipt of your letter of the [*insert date*] in which you purport to confirm an oral direction which you allege was given by [*insert name*] on site/by telephone [*delete as appropriate*].

We hereby formally dissent from such direction in accordance with clause 3.3.3 [*substitute "3.3" when using DSC/C or "4.4" when using DOM/2*] and state that the alleged oral instruction was not given as alleged or at all.

Yours faithfully

函件 186
致指定分包商，附上建筑师关于授权条款的信函
本函件仅适用于使用 NSC/C 合同

Letter 186
To nominated sub-contractor, enclosing architect's letter specifying empowering provision
This letter is only suitable for use with NSC/C

尊敬的先生：

就你方[填入日期]来函中要求我方请建筑师解释主合同条款中关于建筑师于[填入日期]签发的编号为[填入编号]的指令的授权。

现附上我方从建筑师处收到的[填入日期]的来函副本。

你忠诚的

Dear Sir

We refer to your letter of the [*insert date*] in which you required us to request the architect to specify the provision of the main contract empowering the issue of architect's instruction number [*insert number*] dated [*insert date*].

We enclose a copy of a letter dated [*insert date*] which we have received from the architect.

Yours faithfully

函件 187
致建筑师,关于指定分包商工期延误的通知
本函件仅适用于使用 JCT 98 或 NSC/C 合同

Letter 187
To architect, informing him of nominated sub-contractor's notice of delay
This letter is only suitable for use with JCT 98 or NSC/C

尊敬的先生:

  今天我方收到[填入姓名]有关分包工程工期延误的实际情况的通知以及根据分包合同第 2.2.1 条所提供的工期延误细节及估算,请见附件。我方希望能收到你方的认可,包括根据分包合同第 2.3 条你方公正合理地估算的分包工程竣工日期的调整。请告知你方已考虑了哪些因素,包括哪些事件,以及这些事件的影响程度(如有的话),并请告知你方已把这类事件看作要求变更的指令,以及将某部分工作取消、责任取消或限制取消的指令。

  我方希望尽快收到你方的认可,以便我方能将调整的工期控制在 12 周的时间范围内。

你忠诚的

Dear Sir

We have today received a notice from [*insert name*] of the material circumstances of a delay in the sub-contract works together with particulars and estimate in accordance with clause 2.2.1 of the sub-contract. All this information is enclosed herewith and we should be pleased to receive your written consent including the revised period for the completion of the sub-contract work as you estimate to be fair and reasonable in accordance with clause 2.3 of the sub-contract. Please let us know which of the matters, including any of the relevant events, you have taken into account and the extent, if any, to which you have had regard to any instruction requiring as a variation the omission of any work, obligation or restriction.

We should be pleased to receive your consent in good time for us to fix the revised period within the 12 week time limit.

Yours faithfully

函件 188a
致指定分包商,同意工期顺延
本函件仅适用于使用 NSC/C 合同

Letter 188a
To nominated sub-contractor, granting extension of time
This letter is only suitable for use with NSC/C

尊敬的先生：

就你方于[填入日期,若合适,可加入:]发来的工期延误的通知以及你方于[填入日期]发来的进一步信息。

根据分包合同第2.3条并获得建筑师的认可后,我方在此同意工期顺延[填入时间期限],现在分包工程的施工工期调整为[填入时间期限],并于[填入日期]竣工。

已考虑的相关事件为:[列出事件]

我方已考虑以下,作为要求工程变更和取消工程的指令和指示:[列出内容并在工期顺延的时间中扣除]

你忠诚的

抄送:建筑师

Dear Sir

We refer to your notice of delay dated [*insert date and if appropriate add:*] and the further information provided in your letter dated [*insert date*].

In accordance with clause 2.3 of the sub-contract and with the consent of the architect we hereby grant you an extension of time of [*insert period*]. The revised period for completion of the sub-contract works is now [*insert period*] ending on [*insert date*].

The relevant events taken into account are: [*list*]

We have had regard to the following instructions and directions requiring as a variation the omission of work: [*list and include the extent of any reduction in extension of time*]

Yours faithfully
Copy: Architect

函件 188b

致指定分包商，同意工期顺延

本函件仅适用于使用 NAM/SC、DSC/C、DOM/2 或 GC/Works/SC 合同

Letter 188b

To sub-contractor, granting extension of time

This letter is only suitable for use with NAM/SC, DSC/C, DOM/2 or GC/Works/SC

尊敬的先生：

  就你方于[填入日期，若合适，可加入:]发来的工期延误的通知以及你方于[填入日期]发来的进一步信息。

  根据分包合同第 12.2/12.3 条 [视情况取舍，当使用 DSC/C 合同时，为"第 2.3 条"；当使用 DOM/2 合同时，为"第 11.3 条"或使用 GC/Works/SC 合同时，为"第 11.5 条"]我方在此同意你方的分包工程的施工期延长[填入时间期限]。因此，分包工程的施工期调整为[填入时间期限]。

  考虑的事件为：[根据第 12.7、2.10 或 11.10 条列出事件及承包商的违约和每一事件所允许的时间顺延；当使用 GC/Works/SC 合同时，将不适合处取消]

  [当使用 DSC/C 或 DOM/2 合同时，应加入:]

  我方已考虑了下列要求作为工程变更、工程取消的指令：[列出内容并在工期顺延的时间中扣除]

  [当使用 SC/Works/SC，应加入:]

  这是期中的/最终的[视情况取舍]决定。

你忠诚的

Dear Sir

We refer to your notice of delay dated [*insert date and if appropriate add*:] and the further information provided in your letter of the [*insert date*].

In accordance with clause 12.2/12.3 [*delete as appropriate or substitute* "2.3" *when using DSC/C*, "11.3" *when using DOM/2 or* "11.5" *when using GC/Works/SC*] of the sub-contract, we hereby grant you an extension of the period for completion of the sub-contract works of [*insert period*]. The revised period for completion of the sub-contract works is now [*insert period*].

The events taken into account are: [*list events in clause 12.7, 2.10 or 11.10 as appropriate and any default of the contractor and the periods granted in respect of each, but omit where GC/Works/SC is used*].

[*When using DSC/C or DOM/2, add*:]

We have had regard to the following directions requiring as a variation the omission of work: [*list and include the extent of any reduction in extension of time*].

[*When using SC/Works/SC, add*:]

This is an interim/final [*delete as appropriate*] decision.

Yours faithfully

函件 189

当不同意工期顺延时,致分包商

本函件仅适用于使用 NAM/SC、DSC/C、DOM/2 或 GC/Works/SC 合同

Letter 189

To sub-contractor, if no extension of time due

This letter is only suitable for use with NAM/SC, DSC/C, DOM/2 or GC/Works/SC

尊敬的先生:

我方十分认真地审查了你方于[填入日期]发来的关于工期延误的通知[使用 GC/Works/SC 合同时,代之以"请求工期顺延"]以及所附的详细资料。我方认为,在这种情况下,你方无权获得工期顺延/工期的进一步顺延[视情况取舍]。

你忠诚的

Dear Sir

We have carefully examined your notice of delay [*substitute "requesting an extension of time" when using GC/Works/SC*] and accompanying particulars dated [*insert date*] and it is our opinion that you are not entitled to an extension of time/further extension of time [*delete as appropriate*] on this occasion.

Yours faithfully

函件 190a
当工期顺延的索赔无效时,致指定分包商
本函件仅适用于使用 NSC/C 合同

Letter 190a
To nominated sub-contractor, if claim for extension of time is not valid
This letter is only suitable for use with NSC/C

尊敬的先生:

　　收到你方于[填入日期]发来的关于工程延误的通知以及要求顺延工期的证明材料,我方已将你方来函立即转交建筑师,请他根据主合同第35.14条和分包合同第2.2.1条进行考虑。

　　建筑师不同意你方要求顺延工期的请求,现将建筑师于[填入日期]的来函副本附上供参考。

　　[若建筑师的来函中有些内容承包商不愿让分包商看到,最后一段应取消并用下一段的内容替代:]

　　建筑师不同意你方要求顺延工期的请求,他来函的有关内容如下:[填入相关内容]

你忠诚的

Dear Sir

We received your notice of delay and supporting information claiming an extension of time on the [*insert date*] and immediately passed it to the architect for his consideration under the provisions of clause 2.2.1 of the sub-contract and clause 35.14 of the main contract.
The architect will not give his consent to an extension of time and we enclose a copy of his letter dated [*insert date*] for your information.
[*If the architect's letter contains something which the contractor does not wish the sub-contractor to see, the last paragraph should be omitted and the following substituted*:]
The architect will not give his consent to an extension of time. The relevant portion of his letter states: [*insert an exact quotation of the relevant part*].

Yours faithfully

函件 190b
当工期顺延的索赔无效时,致分包商
本函件仅适用于使用 NAM/SC、DSC/C 或 DOM/2 合同

Letter 190b
To sub-contractor, if claim for extension of time is not valid
This letter is only suitable for use with NAM/SC, DSC/C or DOM/2

尊敬的先生:

　　就你方于[填入日期,若合适,请加入:]关于工期延误[当使用 GC/Works/SC 合同时,代之以"请求工期顺延"]的通知以及于[填入日期]来函所提供的进一步资料。

　　基于你方提交的文件资料,我方认为没有理由可顺延工期。我方将乐意考虑你方按分包合同条件并以适当的格式提交进一步资料。

你忠诚的

Dear Sir

We refer to your notice of delay [*substitute "requesting an extension of time" when using GC/Works/SC*] dated [*insert date and if appropriate add:*] and the further information provided in your letter dated [*insert date*].

On the basis of the documents you have presented to us, we see no ground for any extension of time. We shall be pleased to consider any further submissions if they are presented in the proper form and in accordance with the terms of the sub-contract.

Yours faithfully

函件 191
当指定分包商不能按时竣工时,致建筑师
本函件仅适用于使用 JCT 98 合同

Letter 191
To architect, if nominated sub-contractor fails to complete in due time
This letter is only suitable for use with JCT 98

尊敬的先生:

　　[下列两段中任选一段:]

　　[填入指定分包商姓名]已不能在分包合同中规定的工期内完成分包工程[填入工程内容]。

　　[或]

　　[填入指定分包商姓名]已不能在你方认可及我方承诺的工期调整至[填入日期]内完成分包工程[填入工程内容]。

　　[然后加入:]

　　工程施工期于[填入日期]结束。希望你方能按照合同条件第 35.15 条的规定签发竣工证书。

你忠诚的

抄送:指定分包商

Dear Sir

[*Either*:]

[*insert name of nominated sub-contractor*] have failed to complete their [*insert nature of work*] sub-contract work within the period specified in the sub-contract.

[*Or*:]

[*insert name of nominated sub-contractor*] have failed to complete their [*insert nature of work*] sub-contract work within the revised period granted by us on the [*insert date*] with your consent.

[*Then add*:]

The period expired on the [*insert date*] . We should be pleased if you would so certify in accordance with the provisions of clause 35.15 of the conditions of contract.

Yours faithfully

Copy: Nominated sub-contractor

函件 192

当建筑师不同意按第 35.15 条规定签发证书时,致建筑师

本函件仅适用于使用 JCT 98 合同

Letter 192

To architect, if he refuses to issue a clause 35.15 certificate

This letter is only suitable for use with JCT 98

尊敬的先生:

收到你方[填入日期]的来函,得知你方不准备按合同条件第 35.15 条的规定签发证书。在我方[填入日期]的函中已告知你方,[填入姓名]不能按时完成其承担的分包工程。你方不愿签发证书的惟一合法理由是第 35.14 条的要求没有得到很好贯彻。

我方已于[填入日期]将分包工期延误的通知,具体细节及工期延误的估算等转交你方。我方已承诺分包工程施工期顺延至[填入日期]并得到你方的认可[若不合适,可删除]。因此,第 35.14 条的要求已得到了贯彻。若根据第 35.15 条规定,到[填入日期]我方仍未收到你方签发的证书,我方有意立即将此争端提交裁决解决。

你忠诚的

抄送:雇主

Dear Sir

We are in receipt of your letter of the [*insert date*], from which we note that you do not propose to issue your certificate as required by clause 35.15 of the conditions of contract. By our letter of the [*insert date*], we notified you of [*insert name*]'s failure to complete the sub-contract works in due time. The only lawful reason for your failure to so certify would be that clause 35.14 had not been properly applied.

The sub-contractor's notice of delay, particulars and estimates were passed to you by us on the [*insert date*]. We granted an extension of the sub-contract period on the [*insert date*] with your consent [*delete if not applicable*]. Clause 35.14, therefore, has been properly applied. If we do not receive your certificate under clause 35.15 by [*insert date*], we intend to refer the dispute to immediate adjudication.

Yours faithfully

Copy: Employer

函件 193

当未能按时完成工程时,致分包商

本函件仅适用于使用 NAM/SC、DSC/C、DOM/2 或 GC/Works/SC 合同

Letter 193

To sub-contractor, if works not complete by due date

This letter is only suitable for use with NAM/SC, DSC/C, DOM/2 or GC/Works/SC

尊敬的先生:

  根据分包合同第 13 条[使用 DSC/C 合同时,为"第 2.11 条";使用 DOM/2 或 GC/Works/SC 合同时,为"第 12.1 条"],在此,特告知你方,分包工程[填入工程内容]未能在[填入日期]的施工工期末/调整后的施工工期末[视情况取舍]完成。

  你方已违约,特提请注意第 14.3 条 [使用 DSC/C 合同时,为"第 2.12 条";使用 DOM/2 或 GC/Works/SC 合同时,为"第 12.2 条"]。我方保留所有要求赔偿的权利。

你忠诚的

Dear Sir

In accordance with clause 13 [*substitute* "2.11" *when using DSC/C or* "12.1" *when using DOM/2 or GC/Works/SC*] of the sub-contract, we hereby give notice that the sub-contract [*insert nature of works*] works were not completed within the period for completion/revised period for completion [*delete as appropriate*] ending on [*insert date*].

You are in breach of the sub-contract and we draw your attention to clause 14.3 [*substitute* "2.12" *when using DSC/C or* "12.2" *when using DOM/2 or GC/Works/SC*]. We reserve all our rights and remedies.

Yours faithfully

函件 194

关于分包工程实际竣工后审核工期顺延要求时,致建筑师
本函件仅适用于使用 JCT 98 或 NSC/C 合同

Letter 194

To architect, regarding review of extensions after practical completion of sub-contract works
This letter is only suitable for use with JCT 98 and NSC/ C

尊敬的先生:

收到你方于[填入日期]签发的分包工程[填入工程内容]已实际竣工的证书。

我方希望收到你方根据分包合同第 2.5 条的书面确认,公正合理地同意分包工程的施工期延长或缩短,或确认以前已认可的施工期。

我方希望及时收到你方的认可,以便我方能在第 2.5 条规定的 12 周内通知 [填入姓名]。

你忠诚的

Dear Sir

We are in receipt of your certificate of practical completion of the sub-contract [*insert nature of works*] works dated [*insert date*].

We should be pleased to receive your written consent in accordance with clause 2.5 of the sub-contract to the fixing of a longer or shorter period for the completion of the sub-contract works as you consider to be fair and reasonable or the confirmation of the period previously fixed.

We should be pleased to have your consent in good time for us to notify [*insert name*] within the 12 week period stipulated in clause 2.5.

Yours faithfully

函件 195

致指定分包商,要求为费用索赔提供证据资料

本函件仅适用于使用 NSC/C 合同

Letter 195

To nominated sub-contractor, requesting further information in support of a financial claim

This letter is only suitable for use with NSC/C

尊敬的先生：

　　就你方于[填入日期]的来函要求索赔损失和(或)费用事宜。

　　建筑师已要求我方向你方索取进一步资料以利于建筑师能合理地执行主合同第 26.4 条。

　　根据分包合同第 4.38.1.2 条,我方十分希望获得下列有关资料：

[列出所要的资料]

你忠诚的

Dear Sir

We refer to your claim for loss and/or expense dated [*insert date*].

The architect has requested us to obtain from you further information in order reasonably to enable him to operate clause 26.4 of the main contract.

In accordance with clause 4.38.1.2 of the sub-contract, we should be pleased to receive the following information:

[*List the information required*]

Yours faithfully

函件 196

致指定分包商,要求提供损失和(或)费用的详细资料

本函件仅适用于使用 NSC/C 合同

Letter 196

To nominated sub-contractor, requesting details of loss and/or expense

This letter is only suitable for use with NSC/C

尊敬的先生:

关于你方[填入日期]提交的损失和(或)费用的索赔函和你方[填入日期]提供的进一步资料,建筑师/工料测量师[视情况取舍]已要求我方从你方处获得这些损失和(或)费用的详细资料,以便根据主合同第 26.4 条对这些损失和(或)费用进行合理的确定。

因此要求你方根据分包合同第 4.38.1.3 条提供这些详细资料。

你忠诚的

Dear Sir

We refer to your claim for loss and/or expense dated [*insert date*] and the further information which you submitted to us on the [*insert date*].

The architect/quantity surveyor [*delete as appropriate*] has requested us to obtain from you details of loss and/or expense in order reasonably to enable the ascertainment of that loss and/or expense under clause 26.4 of the main contract.

We hereby request such details in accordance with clause 4.38.1.3 of the sub-contract.

Yours faithfully

函件 197
致分包商,要求提供费用索赔的进一步资料

本函件仅适用于使用 NAM/SC、DSC/C、DOM/2 或 GC/Works/SC 合同

Letter 197
To sub-contractor, requesting further information in support of a financial claim

This letter is only suitable for use with NAM/SC, DSC/C, DOM/2 or GC/Works/SC

尊敬的先生:

  本函谨提及你方[填入日期]提交的损失和(或)费用的索赔函。

  为了使我方能合理地执行分包合同第 14.1 条[使用 DSC/C 合同时,为"第 4.24"条;使用 DOM/2 或 GC/Works/SC 合同时,为"第 13.1 条"],我方十分希望收到下列资料:

[列出所要资料]

你忠诚的

Dear Sir

We refer to your claim for loss and/or expense dated [*insert date*].

In order reasonably to enable us to operate the provisions of clause 14.1 [*substitute* "4.24" *when using DSC/C or* "13.1" *when using DOM/2 or GC/Works/SC*] of the sub-contract, we should be pleased to receive the following information:

[*List the information required*]

Yours faithfully

函件 198

致分包商,要求偿付所蒙受的损失和(或)费用
本函件仅适用于使用 NSC/C、NAM/SC、DSC/C 或 DOM/2 合同

Letter 198

To sub-contractor, applying for payment of loss and/or expense
This letter is only suitable for use with NSC/C, NAM/SC, DSC/C or DOM/2

尊敬的先生:

我方在此根据分包合同第 4.40 条 [使用 NAM/SC 合同时,为"第 14.3 条";使用 DSC/C 合同时,为"第 4.27 条";使用 DOM/2 合同时,为"第 13.4 条"] 发出下列通知和申请:

由于正常的工程施工进度因 [具体描述原因] 而确实受到影响,致使我方蒙受直接损失和(或)费用。

随函附上直接损失和(或)费用的详细计算,十分希望你方能同意这笔总数为 [填入金额] 的金额。

你忠诚的

Dear Sir

We hereby give notice and make application under clause 4.40 [*substitute "14.3" when using NAM/SC, "4.27" when using DSC/C or "13.4" when using DOM/2*] of the sub-contract as follows:

We have been caused direct loss and/or expense because the regular progress of the works has been materially affected by [*describe*].

Particulars of the calculation of such direct loss and/or expense are enclosed and we should be pleased to have your agreement to the amount of [*insert amount*].

Yours faithfully

函件 199
致分包商,关于期中付款通知

Letter 199
To sub-contractor, giving notice of an interim payment

尊敬的先生:

　　本书面通知是关于即将提交的期中付款的价款总额,即[填入金额]。该总额是基于下列依据计算的[填入计算方法,若有证明,请说明]。

你忠诚的

Dear Sir

This is a written notice specifying the amount of interim payment which is proposed to be made, namely: [insert amount]. The amount is calculated on the following basis: [insert the way in which the amount is calculated. If it is by reference to a certificate, so state].

Yours faithfully

函件 200
致分包商，发出不支付工程款的通知

Letter 200
To sub-contractor, giving withholding notice

尊敬的先生：

　　根据分包合同第 4.16.1.2 条 [使用 NAM/SC 合同时，为"第 19.2.4 条"；使用 DSC/C 合同时，为"第 4.16.3 条"；使用 DOM/2 合同时，为"第 21.3.3 条"；使用 GC/Works/SC 合同时，为"第 21.2.3 条"]，本函作为书面通知，即不支付你方于[填入日期]要求支付的工程款总额中的一笔款额为[填入金额]的工程款。

　　不支付的原因以及不支付的款额细节如下：

[列出原因及每一原因所对应的款额，应尽可能详细列出]

你忠诚的

Dear Sir

Take this as written notice in accordance with clause 4.16.1.2 [*substitute "19.2.4" when using NAM/SC, "4.16.3" when using DSC/C, "21.3.3" when using DOM/2 or "21.2.3" when using GC/Works/SC*] that we propose to withhold the sum of [*insert amount*] from the amount notified for payment on [*insert date*].

The grounds for withholding and the amount attributable to each ground are as follows:
[*list the grounds with the amount stated for each ground. The list should be as detailed as possible*].

Yours faithfully

函件 201a
致指定分包商,并附上付款支票
本函件仅适用于使用 NSC/C 合同

Letter 201a
To nominated sub-contractor, enclosing payment
This letter is only suitable for use with NSC/C

尊敬的先生:

　　建筑师于[填入日期]签发了第[填入编号]号期中付款凭证,应付款中包含了分包工程[填入工程内容]的工程款,总额为[填入金额]。毫无疑问,你方已收到建筑师付款指令的副本。在此我方附上总额为[填入金额]的支票,即是扣除你方已同意的2%的现金折扣后你方的应收款。

　　[若合适,请加入:]

　　请注意我方已从你方应收款中扣除了[填入金额]的款额。这笔扣款的计算细节已于[填入日期]送交你方。

　　[再加入:]

　　根据分包合同第4.16.1.1条,我方希望你方能在所附回执上签字并寄回我方作为已付款的证据。

你忠诚的

Dear Sir

The architect issued interim certificate number [*insert number*] on the [*insert date*] and the amount stated as due therein included the sum of [*insert amount*] in respect of the sub-contract [*insert nature of works*] works. No doubt you have already received a copy of the architect's direction to that effect. We enclose our cheque in the sum of [*insert amount*] which represents the amount due to you less the permitted $2\frac{1}{2}\%$ cash discount.

[*If appropriate, insert:*]

Note that we have withheld the sum of [*insert amount*] from money otherwise due to you. Details of the calculation of such sum were sent to you on [*insert date*].

[*Then add:*]

We should be pleased if you would sign and return the enclosed receipt as proof of discharge in accordance with clause 4.16.1.1 of the sub-contract.

Yours faithfully

函件 201b

致分包商,并附上付款支票

本函件仅适用于使用 NAM/SC、DSC/C、DOM/2 或 GC/Work/SC 合同

Letter 201b

To sub-contractor, enclosing payment

This letter is only suitable for use with NAM/SC, DSC/C, DOM/2 or GC/Works/SC

尊敬的先生:

根据分包合同第19.2条[使用 DSC/C 合同时为"第4.15条";使用 DOM/2 合同时为"第21.2条";使用 GC/Work/SC 合同时为"第21.1.1条"],我方附上付款支票,总额为[填入数额]。这是我方于[填入日期]到期的[填入编号]号期中付款,此款额按所附的清单计算得出。

[若合适,请加入:]

请注意我方已从你方应收款中扣除[填入数额]。扣款的计算细节已于[填入日期]送交你方。

你忠诚的

Dear Sir

In accordance with clause 19.2 [*substitute* "*4.15*" *when using DSC/C,* "*21.2*" *when using DOM/2 or* "*21.1.1*" *when using GC/Works/SC*] of the sub-contract, we enclose our cheque in the sum of [*insert amount*]. This is our interim payment number [*insert number*] due on [*insert date*] and calculated as indicated on the enclosed statement.

[*If appropriate, add*:]

Note that we have set-off the sum of [*insert amount*] from money otherwise due to you. Details of the calculation of such sum were sent to you on [*insert date*].

Yours faithfully

函件 202

致建筑师,附上已支付指定分包商工程款的证据

本函件仅适用于使用 JCT 98 或 NSC/C 合同

Letter 202

To architect, enclosing proof of payment to nominated sub-contractor

This letter is only suitable for use with JCT 98 or NSC/C

尊敬的先生:

根据合同条款第 35.13.3 条,现附上[填入姓名]的收据作为我方按第 35.13.2 条规定已支付工程款的证明。

你忠诚的

Dear Sir

In accordance with clause 35.13.3 of the conditions of contract, we enclose a receipt from [*insert name*] as proof of payment to them by us as required under clause 35.13.2.

Yours faithfully

函件 203
当承包商不能提供已支付指定分包商工程款的证明时,致建筑师
本函件仅适用于使用 JCT 98 合同

Letter 203
To architect, if contractor is unable to provide proof of payment to nominated sub-contractor
This letter is only suitable for use with JCT 98

尊敬的先生:

　　根据你方[填入日期]的指令,按合同条件第35.13.2条规定,我方已于[填入日期]支付给[填入姓名]一笔工程款总额为[填入数额]。尽管我方于[填入日期]发出催单,该公司至今仍未寄回我方收据凭证,现附上我方的催单通知的副本。

　　现随函附上我方银行提供的已支付支票的副本,我方相信你方能把该支票副本看作符合合同条件第35.13.3条的规定的证明。

你忠诚的

Dear Sir

In compliance with your direction dated [*insert date*], we discharged the sum of [*insert amount*] on [*insert date*] in favour of [*insert name*] as required by clause 35.13.2. This firm has not returned our form of receipt despite reminders sent on [*insert dates*], copies of which are enclosed.

Our bank has provided us with a copy of the processed cheque which we enclose with this letter. We trust that you will accept this as reasonable proof as required under the provisions of clause 35.13.3.

Yours faithfully

函件 204
致裁决人，并附上书面说明
本函件仅适用于使用 NSC/C、NAM/SC、DSC/C、DOM/2 或 GC/Works/SC 合同
专递/挂号邮件
Letter 204
To adjudicator, enclosing written statement
This letter is only suitable for use with NSC/C, NAM/SC, DSC/C, DOM/2 or GC/Works/SC
Special/recorded delivery

尊敬的先生：

　　根据我方于[填入日期]就分包工程[填入工程内容]的施工与[填入姓名]所签订的分包合同条件第9A.5.2条[使用NAM/SC合同时，为"第35A.5.2条"；使用DOM/2合同时，为"第38A.5.2条"或使用GC/Works/SC合同时，为"第38A.4条"]，我方现附上书面说明，作为对由[填入姓名]于[填入日期]提交你方的争端事件的抗辩细节资料。

你忠诚的

抄送：分包商

Dear Sir

Under the provisions of clause 9A.5.2 [*substitute* "*35A.5.2*" *when using NAM/SC*, "*38A.5.2*" *when using DOM/2 or* "*38A.4*" *when using GC/Works/SC*] of our sub-contract dated [*insert date*] with [*insert name*] for the execution of the sub-contract [*insert nature of works*] works, we enclose a written statement setting out brief particulars of our defence to the referral dated [*insert date*] and put forward by [*insert name*].

Yours faithfully

Copy: Sub-contractor

函件 205
当裁决人已聘用但并无争端时，致分包商
专递邮件和传真件

Letter 205
To sub-contractor if adjudicator appointed, but there is no dispute
Special delivery and fax

尊敬的先生：

我方注意到你方已聘用了一位裁决人。

你方来函表示有意将争端通过裁决解决，但并未提及任何争端或争议需提交裁决去解决。因此，裁决人目前并无裁决权限。我方请你方收回裁决人的聘用并通知其本人。若你方不照此执行，请将此信看作通知，即我方保留所有的权利，在无损于我方利益下，我方将抵制任何我方参与的所谓的裁决决定的实施。

你忠诚的

抄送：裁决人

Dear Sir

We note that you have sought the appointment of an adjudicator.

Your notice of intention to refer to adjudication contains no reference to a dispute or difference capable of being referred to adjudication. Therefore, the adjudicator has no jurisdiction. We invite you to withdraw and inform the adjudicator of the position. If you fail to do so, take this as notice that we will reserve all our rights and our participation in the purported adjudication will be without prejudice to our right to resist the enforcement of any decision.

Yours faithfully

Copy: Adjudicator

函件 206
无争端时,致裁决人
专递邮件和传真件

Letter 206
To adjudicator, if there is no dispute
Special delivery and fax

尊敬的先生:

　　提交方于[填入日期]提交你方的有意将争端通过裁决解决的通知中并无需要裁决的争端事项。因此,你方目前无裁决权限,特请你方放弃受聘。

　　若你方不照此执行,请将此函看作通知,即我方保留所有的权利,在无损于我方的利益下,我方将抵制任何我方参与的所谓的裁决决定的实施并寻求法庭作出你方无权获得任何费用的决定。

　　附上已于[填入日期]发给提交方的信件的副本。

你忠诚的

抄送:分包商

Dear Sir

The notice of intention to refer to adjudication which was submitted by the referring party on [*insert date*] contains no dispute capable of being referred to adjudication. Therefore, you are lacking jurisdiction and we invite you to relinquish your appointment.

If you fail to do so, take this as notice that we will reserve all our rights and our participation in the purported adjudication will be without prejudice to our right to resist the enforcement of any decision and to seek a declaration from the court that you are not entitled to fees.

A copy of the letter dated [*insert date*] which we have sent to the referring party is enclosed.

Yours faithfully

Copy: Sub-contractor

函件 207

当分包商已提前7天错误地发出要暂停履约的意向函时,致分包商

Letter 207

To sub-contractor, if sub-contractor has wrongly sent 7 day notice of intention to suspend performance of obligations

尊敬的先生:

我方已收到你方于[填入日期]的来函,很明显,根据分包合同第4.21.1条[使用NAM/SC合同时为"第19.6条";使用DSC/C合同时为"第4.20条";使用DOM/2合同时为"第21.6条";使用GC/Work/SC合同时为"第24条"]你方有意将来函当作通知发送。

[或加入:]

所谓的通知中有严重错误。

[或:]

我方认为你方所谓的通知中的内容含糊不清。

[或:]

你方所谓的通知中的申辩不正确。

[或:]

你方信中提及的工程款已于[填入日期]支付。

[或:]

不支付工程款的有效通知于[填入日期]发出。

[然后加入:]

因此,你方通知无效,不发生作用。谨告知,若你方进一步暂停合同工程[填入工程

内容]的施工,这类暂停将属不合法,我方保留所有合法权利和要求赔偿的权利。

[当使用 NSC/C 合同时,加:]

按主合同第 35.24.1 条的规定,我方将通知建筑师,根据分包合同第 7.1.1.1 和 7.1.1.2 条的规定,你方已构成违约。

你忠诚的

Dear Sir

We are in receipt of your letter of the [*insert date*] which, apparently, you intend to be a notice in accordance with clause 4.21.1 [*substitute* "19.6" *when using NAM/SC,* "4.20" *when using DSC/C,* "21.6" *when using DOM/2 or* "24" *when using GC/Works/SC*] of the sub-contract.

[*Add either*:]

The purported notice contains a serious error.

[*Or*:]

We are advised that your purported notice is ambiguous.

[*Or*:]

The allegation in your purported notice is incorrect.

[*Or*:]

The payment to which you refer was made on the [*insert date*].

[*continued*]

Letter 207 continued

[*Or:*]

An effective withholding notice was given on the [*insert date*].
[*Then add*:]

Your notice is therefore invalid and of no effect. Take notice that if you suspend further execution of the contract [*insert nature of works*] works, such suspension will be unlawful. We reserve all our legal rights and remedies.

[*When using NSC/C, add*:]

We will inform the architect, as required by clause 35.24.1 of the main contract, that you have made default in respect of the matters referred to in clause 7.1.1.1 and 7.1.1.2 of the sub-contract.

Yours faithfully

函件 208
当分包商提前7天正确地发出要暂停履约的意向函时,致分包商

Letter 208
To sub-contractor, if sub-contractor has correctly sent 7 day notice of intention to suspend performance of obligations

尊敬的先生:

我方已收到你方于[填入日期]的来函,根据分包合同第4.21.1条[使用 NAM/SC 合同时为"第19.6条";使用 DSC/C 合同时为"第4.20条";DOM/2 合同时为"第21.6条"或使用 GC/Work/SC 合同时为"第24条"],该函应看作为通知。

我方对你方认为有必要发出这样的通知而感到遗憾,但我方仍将总额为[填入数额]的支票附上,我方希望你方能接受我方对此疏忽表示的歉意。

[当使用 NSC/C 合同时,应加入:]

我方希望你方能在所附的收据上签收并寄回我方作为根据第4.16.1.1条工程款已支付的证据。

你忠诚的

Dear Sir

We are in receipt of your letter of the [*insert date*] which you have sent as a notice in accordance with clause 4.21.1 [*substitute* "19.6" *when using NAM/SC,* "4.20" *when using DSC/C,* "21.6" *when using DOM/2 or* "24" *when using GC/Works/SC*] of the sub-contract.

We regret that you have felt it necessary to send such a notice, but we are pleased to enclose our cheque in the sum of [*insert amount*]. We hope you will accept our apologies for this oversight.

[*When using NSC/C, add:*]

We should be pleased if you would sign and date the enclosed receipt and return it to us as proof of discharge in accordance with clause 4.16.1.1.

Yours faithfully

函件 209
致分包商,请求调整或记录最终分包合同总价的文件资料

Letter 209
To sub-contractor, requesting documents for adjusting or computing the final sub-contract sum

尊敬的先生:

  根据分包合同第4.23.1.1条[当分包合同 NSC/A 第3.2条适用时为"第4.24.1条";当使用 NAM/SC 合同时为"第19.7.1条";使用 DSC/C 合同并在第4.3条适用时为"第21.7.1条";当使用 DOM/2 合同并在第15.2条合适时为"第21.8.1条"或使用 GC/Work/SC 合同时为"第21.4.1条"],我方十分希望收到所有必要的调整/记录[视情况取舍][使用 NSC/C 合同和 NSC/A 合同并在第3.2条适用时;或使用 DSC/C 合同并在第4.3条适用时;或使用 DOM/2 合同并在第15.2条适用时,请填入"已确认的"]最终分包合同总价的文件资料。

你忠诚的

Dear Sir

In accordance with clause 4.23.1.1 [*substitute* "4.24.1" *when NSC/A article 3.2 applies*, "19.7.1" *when using NAM/SC*, "4.21.1" *when using DSC/C and clause 4.2 applies or* "42.2.1" *when using DSC/C and clause 4.3 applies*, "21.7.1" *when using DOM/2 and clause 15.1 applies or* "21.8.1" *when using DOM/2 and clause 15.2 applies, or* "21.4.1" *when using GC/Works/SC*] of the sub-contract, we should be pleased to receive all documents necessary for the purpose of the adjustment/of computing [*delete as appropriate*] the [*when using NSC/C and NSC/A article 3.2 applies or when using DSC/C and clause 4.3 applies or DOM/2 and clause 15.2 applies, insert* "ascertained"] final sub-contract sum.

Yours faithfully

函件 210
致建筑师,附上指定分包商的文件资料
本函件仅适用于使用 JCT 98 合同

Letter 210
To architect, enclosing nominated sub-contractor's documents
This letter is only suitable for use with JCT 98

尊敬的先生:

根据合同条件第 35.17.2 条,我方附上所有必要关于[填入指定分包商姓名]的分包合同总价的最终调整/已确认的最终分包合同总价的记录 [视情况取舍] 的文件资料。

若你方还需其他资料,请即告知我方。

你忠诚的

Dear Sir

In accordance with clause 35.17.2 of the conditions of contract, we enclose all documents necessary for the final adjustment of the sub-contract sum/the computation of the ascertained final sub-contract sum [*delete as appropriate*] in respect of [*insert name of nominated sub-contractor*].

Please inform us immediately if you require any further information.

Yours faithfully

函件 211

当第 35.17 条的证书在所有分包工程缺陷整改完成前已签发时,致建筑师

本函件仅适用于使用 JCT98 合同

Letter 211

To architect, if clause 35.17 certificate issued before all sub-contract defects remedied

This letter is only suitable for use with JCT 98

尊敬的先生:

我方已收到你方根据合同条件第 35.17 条于[填入日期]所签发的证书。该证书的签发是与第 35.17 条的规定相违背的,因为我方认为[填入姓名]尚未根据分包合同整改好他们有义务整改的所有缺陷、收缩裂缝及其他工程质量问题。我方认为只有这些缺陷全部整改完成才是签发这类证书的前提。

由于你方的违约,我方对现有的、将来可能会产生的任何缺陷、收缩裂缝或其他工程质量问题不承担责任,因此第 35.18 条规定不适用。

你忠诚的

Dear Sir

We are in receipt of your certificate issued under the provisions of clause 35.17 of the conditions of contract and dated [*insert date*]. This certificate is issued in breach of clause 35.17, because in our opinion [*insert name*] have not remedied all the defects, shrinkages and other faults which they are bound to remedy under the sub-contract. Our opinion that such defects have been so remedied is a condition precedent to the issue of such certificate.

In view of your breach, we will take no liability for any defects, shrinkages or other faults which now exist or may appear in the future and the provisions of clause 35.18 cannot apply.

Yours faithfully

函件 212

致分包商,附上最终付款

本函件仅适用于使用 DSC/C 或 DOM/2 合同

Letter 212

To sub-contractor, enclosing final payment

This letter is only suitable for use with DSC/C or DOM/2

尊敬的先生:

我方已收到你方于[填入日期]提交的有关最终付款额的最终报表。建筑师于[填入日期,当使用 DSC/C 合同时,该日期应为此信前 33 天内;或使用 DOM/2 合同时,该日期应为此信前 26 天内]签发了最终付款凭证[当使用 DOM/2 合同时,代之以"关于业主和承包商之间的应付余额,以最终款额和最终报表为准"]。现我方将总额为[填入数额]的支票附上。根据分包合同第 4.23 条[使用 DOM/2 合同时为"第 21.9 条"],支付的这笔总额已反映了最终确认调整的分包合同总价与首次及期中付款之差额。所附的计算依据显示了这笔总额是如何计算得来的。

[若合适,请加入:]

请注意我方已从你方应收款中扣除了[填入数额]款项。我方已于[填入日期]将扣款说明以及扣款计算细节寄于你方。

你忠诚的

Dear Sir

We received the final documentation in connection with your final account on [*insert date*]. The final certificate was issued by the architect [*substitute "final account and final statement became conclusive as to the balance due between the employer and the contractor" when using DOM/2*] on [*insert date which should be no more than 33 days before the date of this letter when using DSC/C or 26 days before the date of this letter when using DOM/2*]. We now enclose our cheque in the sum of [*insert amount*]. This sum, which is paid in accordance with clause 4.23 [*substitute "21.9" when using DOM/2*] of the sub-contract represents the difference between the sub-contract sum as finally adjusted and the first and any interim payments. The enclosed calculations show how we have arrived at such sum.

[*If appropriate, add*:]

Note that we have withheld the sum of [*insert amount*] from money otherwise due to you. A statement of our grounds for so doing and details of the manner in which the amount has been quantified have already been sent to you on [*insert date*].

Yours faithfully

函件 213a
致建筑师,通知他指定分包商的违约
本函件仅适用于使用 JCT 98 合同

Letter 213a
To architect, notifying him of the nominated sub-contractor's default
This letter is only suitable for use with JCT 98

尊敬的先生:

根据合同条件第 35.24.1 条,特通知你方,我方认为,[填入姓名]已对分包合同第 7.1.1/7.1.2/7.1.3/7.1.4 条[视情况取舍]构成违约[填入违约内容]。[若合适,请加入:]现附上表明分包商意见的于[填入日期]来函的副本。

若你方确认分包商已违约,我方希望你方按照合同条件第 35.24.6 条执行。

你忠诚的

Dear Sir

In accordance with clause 35.24.1 of the conditions of contract, we hereby notify you that, in our opinion, [*insert name*] has made default in [*insert nature of default*], being a matter referred to in clause 7.1.1/7.1.2/7.1.3/7.1.4 [*delete as appropriate*] of the sub-contract. [*If appropriate, add:*] The sub-contractor's observations are contained in a letter dated [*insert date*], which we enclose herewith.

If you are reasonably of the opinion that the sub-contractor has made such default, we should be pleased if you would operate the provisions of clause 35.24.6.

Yours faithfully

函件 213b
致建筑师,通知他指定分包商的违约
本函件仅适用于使用 WCD 98 合同

Letter 213b
To architect, notifying him of the named sub-contractor's default
This letter is only suitable for use with WCD 98

尊敬的先生:

　　谨通知你方,我方认为,[填入姓名]已对分包合同[填入分包合同标准格式名称]构成违约[填入违约内容][填入合适的分包合同的合同条款编号以提供此定论的依据]。现附上该分包合同条件的副本供参考。

　　按补充条件第 S4 条,由于[填入姓名]是指定分包商,特请求你方根据第 S4.4.1 条同意我方终止对其雇用的意向。

你忠诚的

Dear Sir

We hereby notify you that, in our opinion, [*insert name*] has made default in [*insert nature of default*] being a matter referred to in clause [*insert clause number as appropriate to the particular sub-contract giving grounds for determination*] of the sub-contract [*insert name of sub-contract form*]. A copy of the clause is attached for your convenience.

Because [*insert name*] is a named sub-contractor under supplementary provision S4, we request your consent under provision S4.4.1 to our intention to determine his employment.

Yours faithfully

函件 214
当指定分包商不再违约时,致建筑师
本函件仅适用于使用 JCT98 合同

Letter 214
To architect if nominated sub-contractor discontinues default
This letter is only suitable for use with JCT 98

尊敬的先生:

根据你方[填入编号]的指令,我方已将违约通知递交[填入姓名],他们已于[填入日期]收到该通知。

按分包合同第 7.1 条允许的 14 天内,其违约行为已终止。

你忠诚的

Dear Sir

In accordance with your instruction number [*insert number*], we issued a notice of default to [*insert name*], which they received on [*insert date*].

The default was discontinued within the 14 day period allowed by the sub-contract clause 7.1.

Yours faithfully

函件 215

致分包商,在合同终止前通知其违约

本函件仅适用于使用 NSC/C、NAM/SC、DSC/C 或 DOM/2 合同

专递/挂号邮件

Letter 215

To sub-contractor, giving notice of default before determination

This letter is only suitable for use with NSC/C, NAM/SC, DSC/C or DOM/2

Special/recorded delivery

尊敬的先生:

根据分包合同第 7.1 条[当使用 NAM/SC 合同时,为"第 27.1 条"或当使用 DOM/2 时,为"第 29.2 条"],特通知你方已在以下方面违约:

[合适时,填入违约的时间及细节]

若你方收到此通知的 14 天内 [使用 NAM/SC、DSC/C 或 DOM/2 合同时,为"10 天"]继续违约,或你方在任何时候再次违约,且不管以前是否违约过,我方将根据分包合同,在继续违约或重复违约的 10 天之内 [使用 NSC/C 合同时,可填入"请建筑师作出指示"]决定是否终止对你方的雇用。

你忠诚的

抄送:建筑师[使用 NSC/C 合同时适用]

Dear Sir

We hereby give you notice under clause 7.1 [*substitute* "*27.1*" *when using NAM/SC or* "*29.2*" *when using DOM/2*] of the sub-contract that you are in default in the following respect:

[*Insert details of the default with dates if appropriate*]

If you continue the default for 14 [*substitute* "*10*" *when using NAM/SC, DSC/C or DOM/2*] days after receipt of this notice or if you at any time repeat such default, whether previously repeated or not, we may within 10 days of such continuance or repetition [*when using NSC/C insert:* "*apply to the architect for an instruction to*" *if appropriate*] determine your employment under this sub-contract forthwith.

Yours faithfully

Copy: Architect [*when using NSC/C*]

函件 216
当指定分包商继续违约时,致建筑师
本函件仅适用于使用 JCT 98 合同

Letter 216
To architect, if nominated sub-contractor continues default
This letter is only suitable for use with JCT 98

尊敬的先生:

根据你方[填入编号]的指令,我方已将违约通知递交[填入姓名],他们已于[填入日期]收到该通知。

分包合同第7.1条允许的14天期限已于昨天到期,在整个期限内,[填入姓名]继续有违约行为。

若我方准备根据分包合同在允许的10天期限内发出终止对其雇用的通知,根据合同条件第35.24.6.1条,特请你方就此时事立即发出进一步指令。

你忠诚的

Dear Sir

In accordance with your instruction number [*insert number*], we issued a notice of default to [*insert name*], which they received on [*insert date*].

The 14 days allowed by sub-contract clause 7.1 expired yesterday and, during the whole of the period, [*insert name*] continued their default.

If we are to issue a notice determining their employment under the sub-contract within the 10 days allowed, we require your immediate further instruction to that effect in accordance with clause 35.24.6.1 of the conditions of contract.

Yours faithfully

函件 217
当有可能终止对指定人员的雇用时,致建筑师
本函件仅适用于使用 IFC 98 合同

Letter 217
To architect, if determination of named person's employment possible
This letter is only suitable for use with IFC 98

尊敬的先生:

  根据合同条件第 3.3.3 条的规定,我方必须提醒你方,下列事件有可能导致对 [填入姓名] 雇用的终止。

  [描述事件内容]
我方希望收到你方的指令。

你忠诚的

Dear Sir

In accordance with clause 3.3.3 of the conditions of contract, we have to advise you that the following events are likely to lead to the determination of [*insert name*]'s employment:

[*Describe the events*]

We should be pleased to receive your instructions.

Yours faithfully

函件 218
致建筑师,寻求终止指定分包商的雇用的指令
本函件仅适用于使用 JCT 98 合同

Letter 218
To architect, seeking instructions to determine nominated sub-contractor's employment
This letter is only suitable for use with JCT 98

尊敬的先生:

我方认为,我方有权因为[填入理由]而终止对[填入姓名]的雇用。

根据合同条件第 35.25 条的规定,请你方将终止雇用合同的指令发给我方。我方十分希望你方能尽快重新指定分包商。由于合同终止,有许多具体问题应进行讨论,因而我方建议,若你方认为方便的话,我方将于[填入日期]拜访你方。

你忠诚的

Dear Sir

We consider that we have the right to determine the employment of [*insert name*] because [*insert reasons*].

Please let us have your instruction to determine as provided in clause 35.25 of the conditions of contract. We should be pleased if you would proceed to make a further nomination of sub-contractor as soon as possible. There are a great many detailed matters which we should discuss as a result of the determination and we suggest that we should visit your office on the [*insert date*] if that is convenient.

Yours faithfully

函件 219a

致分包商，在发出违约通知后终止雇用关系

本函件仅适用于使用 NSC/C、NAM/SC、DSC/C 或 DOM/2 合同

专递/挂号邮件

Letter 219a

To sub-contractor, determining employment after default notice

This letter is only suitable for use with NSC/C, NAM/SC, DSC/C or DOM/2

Special/recorded delivery

尊敬的先生：

　　谨此提及我方于[填入日期]发给你方的违约通知。

　　根据第7.1条[当使用NAM/SC合同时，为"第27.1条"；或使用DOM/2合同时，为"第29.2条"]请将本函作为通知，因而我方决定根据分包合同，在此立即终止对你方的雇用，并无损于我方可能拥有的其他任何索赔权利。

　　第7.4条和第7.5条[使用NAM/SC合同时，为"第27.3条"；使用DOM/2合同时，为"第29.5条和第29.6条"]赋予了合同双方的权利和义务。当我方作出这一决定之前，所有临时设施、设备、工具、机具、货物或材料均不得运离现场。

你忠诚的

Dear Sir

We refer to the default notice sent to you on the [*insert date*].

Take this as notice that, in accordance with clause 7.1 [*substitute "27.1" when using NAM/SC or "29.2" when using DOM/2*], we hereby forthwith determine your employment under this sub-contract without prejudice to any other rights or remedies which we may possess.

The rights and duties of the parties are governed by clauses 7.4 and 7.5 [*substitute "clause 27.3" when using NAM/SC or "clauses 29.5 and 29.6" when using DOM/2*]. No temporary buildings, plant, tools, equipment, goods or materials shall be removed from site until and if we so direct.

Yours faithfully

函件 219b
致分包商,终止雇用关系
本函件仅适用于使用 NAM/SC、DSC/C 或 DOM/2 合同
专递/挂号邮件
Letter 219b
To sub-contractor, determining employment
This letter is only suitable for use with NAM/SC, DSC/C or DOM/2
Special/recorded delivery

尊敬的先生:

　　根据分包合同第27.2条[使用 DSC/C 合同时,为"第7.2条",或使用 DOM/2 合同时,为"第29.3条"],请将本函看作通知,根据本分包合同,我方在此立即终止对你方的雇用,并无损于我们可能拥有的其他任何索赔权利。

　　第27.3条[使用 DSC/C 合同时,为"第7.4和7.5条",或使用 DOM/2 合同时,为"第29.5和29.6条"]赋予了合同双方的权利和义务。请注意在分包工程竣工前,我方无义务再向你方支付任何工程款项,我方保留此权利直至工程竣工。

你忠诚的

Dear Sir

In accordance with clause 27.2 [*substitute* "*7.2*" *when using DSC/C or* "*29.3*" *when using DOM/2*] of the sub-contract, take this as notice that we hereby forthwith determine your employment under this sub-contract without prejudice to any other rights and remedies which we may possess.

The rights and duties of the parties are governed by clause 27.3 [*substitute* "*clauses 7.4 and 7.5*" *when using DSC/C or* "*clauses 29.5 and 29.6*" *when using DOM/2*]. Take note that we are not bound to make any further payments to you until after the completion of the sub-contract works. We reserve our rights to that time.

Yours faithfully

函件 220a

当指定分包商继续违约,承包商终止分包合同时,致建筑师

本函件仅适用于使用 JCT98 合同

Letter 220a

To architect, if nominated sub-contractor continues default and contractor determines sub-contract

This letter is only suitable for use with JCT 98

尊敬的先生:

　　根据你方[填入编号]指令,我方已将违约通知递交[填入姓名],他们已于[填入日期]收到该通知。

　　分包合同第7.1条允许的14天期限已于[填入日期]到期,在整个期限内,[填入姓名]继续有违约行为。

　　因此,我方于[填入日期]发出了终止合同的通知。现附上该通知的副本。

　　根据合同条件第35.24.6.3条的规定,特请你方重新指定分包商。由于合同终止,有许多细节需要讨论。若你方认为方便的话,我方计划于[填入日期]拜访你方。

你忠诚的

Dear Sir

In accordance with your instruction number [*insert number*], we issued a notice of default to [*insert name*], which they received on [*insert date*].

The 14 days allowed by sub-contract clause 7.1 expired on the [*insert date*] and, during the whole of the period, [*insert name*] continued their default. Accordingly, we issued a notice of determination on [*insert date*], a copy of which is enclosed.

We should be pleased if you would proceed to make a further nomination in accordance with the provisions of clause 35.24.6.3. There are a great many detailed matters which we should discuss as a result of the determination and we suggest that we should visit your office on the [*insert date*] if that is convenient.

Yours faithfully

函件 220b
当指定人员的雇用被终止时,致建筑师
本函件仅适用于使用 IFC 98 合同

Letter 220b
To architect, if employment of named person determined
This letter is only suitable for use with IFC 98

尊敬的先生:

[若承包商已告知建筑师可能会导致合同终止的事件时,开始:]
就我方于[填入日期]发出的忠告,

[然后,或在其他情况下应开始:]

根据合同条件第3.3.3条,必须告知你方,我方已于[填入日期]终止对[填入姓名]的雇用,其原因是[填入原因,合适时将合同条款编号和日期写上]。
根据第3.3.3条规定,我方希望能收到你方就此发出的指令。

你忠诚的

Dear Sir

[*If the contractor has advised architect of events likely to lead to determination, begin*:]
Further to our advice of the [*insert date*],

[*Then, or otherwise begin*:]

We must notify you in accordance with clause 3.3.3 of the conditions of contract, that the employment of [*insert name*] was determined on the [*insert date*]. The circumstances are that [*insert circumstances, quoting clause numbers and dates as appropriate*].

We should be pleased to receive your instructions in accordance with clause 3.3.3.

Yours faithfully

函件 220c

当根据 NAM/SC 合同第 27.1 条或第 27.2 条已终止对指定人员的雇用时,致建筑师

本函件仅适用于使用 IFC 98 合同

Letter 220c

To architect, if employment of named person determined under NAM/SC clause 27.1 or 27.2

This letter is only suitable for use with IFC 98

尊敬的先生:

根据 NAM/SC 合同条件第 27.1 条/第 27.2 条[视情况取舍],我方于[填入日期]已终止[填入姓名]的 [填入工程内容] 分包合同。根据主合同条件第 3.3.6(b)条,我方已采取合理措施来弥补雇主因执行第 3.3.4(a)条/第 3.3.4(b)条/第 3.3.5 条[视情况取舍] 而支付我方的额外费用以及根据主合同第 2.7 条而不执行第 3.3.4(a)条/第 3.3.4(b)条/第 3.3.5 条[视情况取舍]我方可能需支付雇主的相当于工程延误违约金的一笔费用。这两笔费用的总额为[填入数额],其明细如下:[填入分项明细]。

[另加:]

我方附上 [填入日期] 发给 [填入姓名] 的函件的复印件,我方认为这是合理的行为。

[继续]

函件 220c 继续

[或:]

我方现附上 [填入日期] 发给 [填入姓名] 的函件的副本,我方认为这是合理的行为。同时附上的是[填入姓名]于[填入日期]来函的副本,从中你方可以看出他们对付款的责任提出争议。尽管我方将继续努力争取得到付款,但我方认为这些努力可能会付诸东流。请告知我方,你方是否要求我方对[填入姓名]启动仲裁或法律诉讼程序。只要雇主书面同意免除我方因根据第 3.3.6 条而产生的任何法律费用,我方准备启动这类诉讼程序。

你忠诚的

抄送:雇主

Dear Sir

We refer to the determination of the [*insert nature of work*] sub-contract employment of [*insert name*] on [*insert date*] under the provisions of NAM/SC clause 27.1/27.2 [*delete as appropriate*]. In accordance with the provisions of clause 3.3.6(b) of the conditions of main contract, we have taken reasonable action to recover additional amounts payable to us by the employer as a result of the application of clause 3.3.4(a)/3.3.4(b)/3.3.5 [*delete as appropriate and add, if appropriate*:] together with an amount equal to the liquidated damages that would have been payable or allowable by us to the employer under clause 2.7 of the main contract, but for the application of clause 3.3.4(a)/3.3.4(b)/3.3.5 [*delete as appropriate*]. The total of such amounts is [*insert total*] made up as follows: [*insert a breakdown of the amount*].

[*Add either*:]

We attach copies of our letters of the [*insert dates*] to [*insert name*] which we consider to be reasonable action.

[*Or*:]

We attach copies of our letters dated [*insert dates*] to [*insert name*] which we consider to be reasonable action. Also enclosed is a copy of a letter from [*insert name*] dated [*insert date*] from which you will see that they dispute their liability to pay. Although we are prepared to continue our efforts to secure payment, we do not consider that our efforts are likely to meet with success. Please inform us if you require us to commence arbitration or other proceedings in respect of [*insert name*]. We are prepared to commence such proceedings only if the employer agrees in writing to indemnify us against any legal costs in accordance with the provisions of clause 3.3.6.

Yours faithfully

Copy: Employer

函件 221
当要求承包商批准替代分包商的报价时,致建筑师
本函仅适用于 JCT 98 合同

Letter 221
To architect, if contractor asked to consent to the price of the substituted sub-contractor
This letter is only suitable for use with JCT 98

尊敬的先生:

你方于[填入日期]的来函收悉,我方注意到你方提议指定[填入姓名]作为[填入工程范围]工程的指定分包商并以[填入报价]的价格分包该工程。

[如果同意,可加入:]

根据合同第 35.18.1.2 条,我方同意该报价。

[如果不同意,可加入:]

根据合同第 35.18.1.2 条,我方不同意这个报价,因为[填入理由]。

你忠诚的

Dear Sir

Thank you for your letter of the [*insert date*] in which we note that you propose to nominate [*insert name*] as substituted sub-contractor for [*insert the nature of the works*] at a price of [*insert amount*].

[*If agreeing, add*:]

In accordance with clause 35.18.1.2, we agree to this price.

[*If not agreeing, add*:]

In accordance with clause 35.18.1.2, we do not agree to this price, because [*insert reasons*].

Yours faithfully

函件 222
当分包合同终止后,承包商被指令实施指定人员的工程时,致建筑师
本函件仅适用于使用 IFC 98 合同

Letter 222
To architect, if contractor instructed to carry out named person's work after determination
This letter is only suitable for use with IFC 98

尊敬的先生:

收到你方根据合同条件第 3.3.3(b)条发出的指令。

若我方希望分包的话,该条款允许我方将工程[填入工程名称]进行分包。我方正对此事进行紧急磋商,一旦我方决定分包出去,我方将再次函告你方。

你忠诚的

[注意:第 3.3.3(b)条指令可归属于第 2.3 条规定的事件,第 4.12 条规定的事件及一项变更事项]

Dear Sir

We are in receipt of your instructions under the provisions of clause 3.3.3(b) of the conditions of contract.

This clause permits us to sub-contract the [*insert nature of work*] work if we so wish. We are giving the matter our urgent consideration and if we decide to so sub-let, we will write to you again.

Yours faithfully

[*Note that a clause 3.3.3(b) instruction ranks as a clause 2.3 event, a clause 4.12 matter and as a variation*]

函件 223
当指定人员的雇用已终止,承包商决定分包时,致建筑师
本函件仅适用于使用 IFC 98 合同

Letter 223
To architect, if contractor decides to sub-let after determination of named person's employment
This letter is only suitable for use with IFC 98

尊敬的先生:

　　就我方[填入日期]的函,我方已决定将[填入工程内容]工程分包给[填入分包商姓名及地址]。

　　[若合适时,可加入:]

　　你方可能知晓,我方通常自己不实施这类工程,常分包给[填入分包商姓名],这是最佳的处理方式。但对于工程进度计划也许会有些问题,我们正在努力解决。

你忠诚的

[注意:第 3.3.3(b)条指令可归属于第 2.3 条规定的事件,第 4.12 条规定的事件及一项变更事项]

Dear Sir

Further to our letter of the [*insert date*], we have decided to sub-let the [*insert nature of work*] work to [*insert name and address of the sub-contractor*].

[*If appropriate, add:*]

You may be aware that we do not usually carry out work of this type ourselves and sub-letting to [*insert name of sub-contractor*] represents the best way of dealing with the matter. However, there may be some difficulties with regard to programming which we are trying to resolve.

Yours faithfully

[*Note that a clause 3.3.3(b) instruction ranks as a clause 2.3 event, a clause 4.12 matter and as a variation*]

函件 224
当分包合同终止且费用已收回时,致雇主
本函件仅适用于使用 WCD 98 或 IFC 98 合同

Letter 224
To employer, if money recovered after determination
This letter is only suitable for use with WCD 98 or IFC 98

尊敬的先生:

　　本函就我方[填入日期]发出的关于我方根据补充协议第 S4.4.3 条[当使用 IFC98 合同时,用"合同条款第 3.3.6 条"替代],从[填入姓名]处收回费用所做出的努力。
　　十分高兴地向你方报告,我方代表你方所做出的努力已见成效,已收回的款项为[填入金额],具体如下:[填入分项细节]。

你忠诚的

Dear Sir

We refer to our letter of the [*insert date*] regarding our efforts to recover amounts from [*insert name*] in accordance with supplementary provision S4.4.3 [*substitute "clause 3.3.6 of the conditions of contract" when using IFC 98*].

We are pleased to report that our efforts on your behalf have been successful and we have recovered the sum of [*insert amount*] which is made up as follows: [*insert breakdown of amount*].

Yours faithfully

函件 225
当指定分包商告知工程已实际竣工时,致建筑师
本函件仅适用于使用JCT 98合同

Letter 225
To architect, if nominated sub-contractor notifies practical completion
This letter is only suitable for use with JCT 98

尊敬的先生:

现附上[填入姓名]根据分包合同第2.10条发出的通知,告知分包合同[填入工程内容]将于[填入日期]达到实际竣工。

[若有要求,承包商应填入有关记录。若已填入有关记录,本函的副本应递交指定分包商]若你方能按合同条件第35.16条的规定,签发分包合同工程实际竣工证书,我方将十分感谢。

你忠诚的

Dear Sir

We enclose a notice received from [*insert name*] in accordance with clause 2.10 of the sub-contract to the effect that the sub-contract [*insert nature of the work*] works will have reached practical completion on the [*insert date*].

[*If desired, the contractor should insert any observations. If observations are inserted, a copy of this letter should be sent to the nominated sub-contractor*]

We should be pleased if you would issue your certificate of practical completion of the sub-contract works as required by clause 35.16 of the conditions of contract.

Yours faithfully

函件 226
致指定分包商,附上实际竣工时的记录
本函件仅适用于使用 NSC/C 合同

Letter 226
To nominated sub-contractor, enclosing observations at practical completion
This letter is only suitable for use with NSC/C

尊敬的先生:

  收到你方通知,按照你方意见,分包合同工程[填入工程内容]将于[填入日期]实际竣工。

  你方通知连同我方记录已于今日一并送交建筑师。根据分包合同第2.10条,附上我方记录的副本供参考。

你忠诚的

Dear Sir

We are in receipt of your notification stating that, in your opinion, practical completion of the sub-contract [*insert nature of works*] works will be achieved on [*insert date*].

Your notification has today been passed to the architect together with our observations. In accordance with clause 2.10 of the sub-contract, a copy of our observations is enclosed for your information.

Yours faithfully

函件 227
当承包商不同意分包工程的实际竣工日期时，致分包商
本函件仅适用于使用 NAM/SC、DSC/C、DOM/2 或 GC/Works/SC 合同

Letter 227
To sub-contractor if contractor dissents from date of practical completion
This letter is only suitable for use with NAM/SC, DSC/C, DOM/2 or GC/Works/SC

尊敬的先生：

收到你方于[填入日期]的通知，按照你方意见，分包合同工程[填入工程内容]将于[填入日期]实际竣工[使用 GC/Work/SC 合同时，代之以"竣工"]。

根据分包合同第 15.1 条[使用 DSC/C 合同时，为"第 2.13 条"或使用 DOM/2 合同或 GC/Work/SC 合同时，为"第 14.1 条"]，请将本函看作正式通知，即我方因下列原因不同意你方信中提及的实际竣工日期[使用 GC/Works/SC 合同时，代之以"竣工日"]。

[简单地列出理由]

你忠诚的

Dear Sir

We are in receipt of your notification dated [*insert date*] stating that, in your opinion, practical completion [*substitute "completion" when using GC/Works/SC*] of the sub-contract [*insert nature of works*] works will be achieved on [*insert date*].

Take this as formal notice, in accordance with clause 15.1 [*substitute "2.13" when using DSC/C or "14.1" when using DOM/2 or GC/Works/SC*] of the sub-contract, that we dissent from such date of practical completion [*substitute "completion" when using GC/Works/SC*] for the following reasons:

[*State reasons clearly, but briefly*].

Yours faithfully

函件 228
致分包商,附上缺陷清单

Letter 228
To sub-contractor, enclosing schedule of defects

尊敬的先生:

　　主合同中的缺陷责任期[使用 GC/Works/SC 合同时,代之以"维修期"]于[填入日期]结束,我方已从建筑师处收到一份缺陷清单。

　　根据分包合同第 2.12 条[使用 NAM/SC 合同时,为"第 15.3 条";使用 DSC/C 合同时,为"第 2.15 条";使用 DONM/2 或 GC/Works/SC 合同时,为"第 14.3 条"],部分缺陷属你方的责任,现附上你方应负责的缺陷清单并请进行必要的整改工作。

你忠诚的

Dear Sir

The defects liability [*substitute "maintenance" when using GC/Works/SC*] period in the main contract ended on the [*insert date*] and we have received a schedule of defects from the architect.

Certain of the defects are your responsibility under clause 2.12 [*substitute "15.3" when using NAM/SC, "2.15" when using DSC/C or "14.3" when using DOM/2 or GC/Works/SC*] of the sub-contract and we enclose a schedule of such defects for your attention. Please carry out the necessary remedial work forthwith.

Yours faithfully

函件 229
致分包商,指出某些缺陷尚未整改完成

Letter 229
To sub-contractor, directing that some defects are not to be made good

尊敬的先生:

　　主合同规定的缺陷责任期将于[填入日期]结束,我方已收到建筑师送来的缺陷清单。

　　根据分包合同第 2.12 条[使用 NAM/SC 合同时,为"第 15.3 条",使用 DSC/C 合同时,为"第 2.15 条"或使用 DON/2 或 GC/Works/SC 合同时,为"第 14.3 条"],部分缺陷属你方责任,因此附上一份这类缺陷清单,请整改。

　　[或:]

建筑师指示,缺陷清单上的缺陷不必由你方整改。

　　[或:]

建筑师指示,缺陷清单上标有"E"的缺陷不必由你方整改。

　　[然后:]

有关不必由你方整改的缺陷,将从你方分包合同中扣除一笔适当的款项。

你忠诚的

Dear Sir

The defects liability period in the main contract ended on the [*insert date*] and we have received a schedule of defects from the architect.

Certain of the defects are your liability under clause 2.12 [*substitute* "15.3" *when using NAM/SC*, "2.15" *when using DSC/C or* "14.3" *when using DOM/2 or GC/Works/SC*] of the sub-contract and we enclose a schedule of such defects for your attention.

[*Either*:]

The architect has instructed that you are not required to make good any of the defects shown on the schedule.

[*Or*:]

The architect has instructed that you are not required to make good those defects marked "E".

[*Then*:]

An appropriate deduction will be made from the sub-contract in respect of the defects which you are not required to make good.

Yours faithfully

函件 230
当指定分包商设计失误而使承包商受到威胁时，致建筑师
本函件仅适用于使用 JCT 98 合同

Letter 230
To architect, if action threatened because of nominated sub-contractor's design failure
This letter is only suitable for use with JCT 98

尊敬的先生：

  谢谢你方[填入日期]的来函。
  我方注意到雇主正在设法使我方对由[填入姓名]设计并施工的分包合同工程[填入工程内容]的缺陷承担责任。
  合同条件第35.12条已明确规定我方不承担此类缺陷的责任。

你忠诚的

抄送：雇主

Dear Sir

Thank you for your letter of the [*insert date*].

We note that the employer is seeking to hold us responsible for a defect in the sub-contract [*insert nature of works*] works designed and carried out by [*insert name*].

Our liability in respect of such defect is expressly excluded by clause 35.21 of the conditions of contract.

Yours faithfully

Copy: Employer

函件 231

当指定人员的设计失误而使承包商受到威胁时,致建筑师

本函件仅适用于使用 IFC 98 合同

Letter 231

To architect, if action threatened because of named person's design failure

This letter is only suitable for use with IFC 84

尊敬的先生:

　　谢谢你方[填入日期]的来函。

　　我方注意到雇主正在设法使我方对由[填入姓名]设计并施工的分包合同工程[填入工程内容]的缺陷承担责任。

　　合同条件第 3.3.7 条已明确规定我方不承担此类缺陷的责任。

你忠诚的

抄送:雇主

Dear Sir

Thank you for your letter of the [*insert date*].

We note that the employer is seeking to hold us responsible for a defect in the sub-contract [*insert nature of work*] works designed and carried out by [*insert name*].

Our liability in respect of such defect is expressly excluded by clause 3.3.7 of the conditions of contract.

Yours faithfully

Copy: Employer

函件 232
关于职业责任保险,致分包合同的建筑师、工程师或其他工程顾问
本函件仅适用于使用 WCD 98 合同

Letter 232
To sub-contract architect, engineer or other consultant, regarding professional indemnity insurance
This letter is only suitable for use with WCD 98

尊敬的先生:

  关于想请你方从事本工程项目的部分设计的可能性问题,我们于[填入日期]进行了电话交谈。现我方确认,我方已与雇主[填入姓名]采用 JCT 的设计加施工标准合同文本(即 WCD98 合同)签订了合同。

  我方正在考虑你方于[填入日期]提交的设计费用建议书,但在做出进一步决定前,我方希望收到你方已办理足够且合适的职业责任保险的证明,并希望将此类保险保持[填入期限]有效,直至工程实际竣工。

你忠诚的

Dear Sir

We refer to our telephone conversation of [*insert date*] regarding the possibility that we shall employ you to carry out certain design functions in relation to this project. We confirm that we are engaged under the JCT Standard Form of Building Contract With Contractor's Design (WCD 98) by the employer, [*insert name*].

We are considering the fee proposal you put to us on the [*insert date*] but before we can take the matter further, we should be pleased to receive evidence that you maintain adequate and suitable professional indemnity insurance and that you will continue to maintain such insurance for a period of [*insert period*] after practical completion of the works.

Yours faithfully

函件 233

关于担保事项,致分包合同的建筑师、工程师或其他工程顾问

本函件仅适用于使用 WCD 98 合同

Letter 233

To sub-contract architect, engineer or other consultants, regarding warranties

This letter is only suitable for use with WCD 98

尊敬的先生:

  随信附上两份担保书格式,你方投标资料中附有此格式,且担保书是我们之间所签订合同的一部分。这些担保书文件必须作为契约签字/履行[视情况取舍],然后我方才能与你方签订工程项目设计合同。

  请在两份格式中的指定位置填好并尽快寄回我方。我方将把这两份格式担保书提交雇主去填写完成,并在规定时间内将一份格式担保寄回你方存档。

你忠诚的

Dear Sir

We enclose two copies of the form of warranty which were included in the information on which you tendered and which formed part of the contract between us. The documents must be signed/executed as a deed [*delete as appropriate*] before we can enter into a contract with you to carry out design functions in connection with this project.

Please complete both forms where indicated and return them to us as soon as possible. We shall send them to the employer for completion and return one form to you for your records in due course.

Yours faithfully

函件 234
未能及时提供资料时，致分包合同的建筑师、工程师或其他工程顾问
本函件仅适用于使用 WCD 98 合同

Letter 234
To sub-contract architect, engineer or other consultant, if late in providing information
This letter is only suitable for use with WCD 98

尊敬的先生：

在此通知你方，至今我方尚未收到应由你方在适当的时间期限内提供的资料，致使我方未能下定单/制作构件/实施工程[视情况取舍]。下面的事实反映了这一点：

[将要求提供的图纸内容及日期列表并提供实际提供图纸的日期]

你方应还记得我们曾于工程开工前[填入日期]一起讨论出图问题。你方出图的延误造成了我方费用增加，根据主合同我方也许可能蒙受工期延误损害赔偿费用。当然我方没有理由要求雇主补偿我方的损失。坦率地说，你方不能按我们双方认可的时间表提供资料，实际上是一种违约行为，我方有权获得赔偿。目前，就此事我方保留所有的权利和索赔要求。

请注意，关键是我方最迟应在[填入日期]收到所有延误的资料，然后在以后的日子里继续收到进一步资料。若今后再发生资料提供延误，我方所蒙受的损失和增加的费用将从你方应收款或将成为你方应收款中扣除。

你忠诚的

Dear Sir

This is to inform you that we are not receiving the information, which it is your responsibility to prepare, within the appropriate time periods to enable us to place our orders/make the components/carry out the work [*delete as appropriate*]. The following schedule speaks for itself:

[*List the descriptions of drawings requested with dates and the dates actually supplied*]

You will recall that we jointly prepared the drawing issue schedule on [*insert date*] before work began on site. Your delays are causing us expense and possibly we may be liable for liquidated damages under the main contract. We certainly have no grounds for recovery of our loss from the employer. To be frank, your delay in providing us with information in conformity with the jointly agreed schedule is a breach of contract for which we are entitled to recover damages. For the moment we are reserving all our rights and remedies in that regard. Take note that it is essential that we receive all the delayed information by [*insert date*] at latest and then continue to receive further information on the due dates. If there are any subsequent delays in receipt of information, we shall set-off our loss and expense against any money due or to become due to you.

Yours faithfully

函件 235
当设计失误而使承包商受到威胁时,致分包合同的建筑师、工程师或其他工程顾问
本函件仅适用于使用 WCD 98 合同

Letter 235
To sub-contract architect, engineer or other consultant, if action threatened because of design failure
This letter is only suitable for use with WCD 98

尊敬的先生:

  我方已收到雇主于[填入日期]的来函,现附上副本。你方可注意到雇主正在设法让我方对工程中的缺陷负责。显然,该缺陷是由于设计失误造成的。

  这部分工程属于你方责任范围,我方已安排[填入日期和时间]在现场/我方办公室[视情况取舍]来讨论该问题并寻求解决方案。我方希望得到你方参加会议的确认。

你忠诚的

Dear Sir

We have received a letter from the employer dated [*insert date*], a copy of which is enclosed. You will note that he is seeking to hold us responsible for a defect in the works. The defect in question appears to be caused by a failure of design.

This part of the work is your responsibility and we have arranged a meeting on site/at our offices [*delete as appropriate*] on [*insert date*] at [*insert time*] to examine the problem and achieve a solution and we should be pleased to have your confirmation that you will attend.

Yours faithfully

函件 236

当工程项目顺利完成时,致分包合同的建筑师、工程师或其他工程顾问

本函件仅适用于使用 WCD 98 合同

Letter 236

To sub-contract architect, engineer or other consultant, at the end of a successful project

This letter is only suitable for use with WCD 98

尊敬的先生:

  现我方附上总额为 [填入金额] 的本工程项目的最终工程款。

  尽管如此规模的工程常会存在一些问题,但所有工程内容已完工,建筑物外形美观,在功能上也能使雇主满意。雇主对工程的最终成果表示十分满意。

  经雇主同意,我方将建筑师拍摄的照片放入我方宣传资料,当然,我方也向你方深表谢意。我方相信,你方不反对这种拍摄,除非我方获知你方明确表示不愿意。当然,我方会将照片及宣传资料送几份给你方。

  我们双方在此工程项目上合作愉快,并期待着在今后的工程项目上能再次合作。

你忠诚的

Dear Sir

We enclose our final payment in the sum of [*insert amount*] in respect of this project.

Despite the usual problems on this scale of work, all is now complete and the building looks well and, apparently, functions to the employer's satisfaction. He has expressed himself as being extremely pleased with the outcome.

With the employer's permission, we are arranging to photograph the building for use in our publicity material where, of course, we shall give you proper acknowledgement. We presume, unless we hear to the contrary, that you have no objection to this course of action. We shall, of course, send you copies of the photographs and of the publicity material.

We have enjoyed working with you on this project and we look forward to cooperating again on future projects.

Yours faithfully

名词解释

Acceptance,	认可,批准
Adjudication	裁决
Advance payment	预付款
Antiquities	文物
Arbitration	仲裁
Architect, replacement	建筑师,替代人选
Architect instructions	建筑师指令
Ascertainment	确认
Assignment	权益转让
At large, time	工期不固定
Best endeavors	尽最大努力
Bills of quantities	工程量清单
Bond	担保,保函
Breach	违约
Certificates	证书
Change	变更,改变
Clerk of works	工程管理人员
Code of practice	施工规范
Code of Procedure for Single Stage Selective Tendering 1996	1996年单一阶段招投标程序规范
Commencement	开工
Common law claims	依据普通法的索赔
Compliance	一致
Confirmation	确认
Consent	同意
Contract documents	合同文件
Contract Sum Analysis	合同总价分析
Contractor's Proposals	承包商建议书
Contractor's Statement	承包商说明

Covering up	覆盖
Daywork	以日工单价计价的工作
Deed	契约
Default notice	违约通知
Defects	缺陷
Defects liability period	缺陷责任期
Deferment of possession	工地占有日期的延误
Delay	工期延误
Design	设计
Design fault	设计失误
Determination	决心,决定
Development control decision	控制开发项目的决定
Directions	指示,指令
Discrepancies	差异,不一致
Dispute	争端,分歧
Disruption	中断
Divergence	差异,不一致
Drawings	图纸
Emergency compliance	紧急事件
Employer's licensees	雇主许可证
Employer's Requirements	雇主要求
Errors	错误
Excavations	开挖
Expense	费用
Extensions of time	工期顺延时间
Delay	延误
Grounds	原因
Review	审查
Sufficiency	足够
Failure of work	工程不合格
Final account and final statement	最终账单和最终报表
Final certificate	最终证书

Housing Grants, Construction and Regeneration Act 1996
　　　　　　　　　　　　　　　1996年住房补贴,建造和维修法案
Inconsistencies　　　　　　　　不一致
Incorporation　　　　　　　　　包括,包含
Information release schedule　　信息公布时间表

Injunction　　　　　　　　　　强制命令
Insolvency　　　　　　　　　　破产
Inspectin　　　　　　　　　　　检查
Insurance　　　　　　　　　　保险
Certificate　　　　　　　　　　证书
Contractor　　　　　　　　　　承包商
Damage　　　　　　　　　　　损害,损失
Employer　　　　　　　　　　雇主
Failure to maintain　　　　　　未能续保
Professional indemnity　　　　　专业保障
Sub-contractior　　　　　　　　分包商
Interest　　　　　　　　　　　利息,利益

Joint Fire Code　　　　　　　　联合消防规范
Jurisdiction　　　　　　　　　司法权,管辖范围
Liquidated damages　　　　　　工期延误损害费
Loss and/or expenses　　　　　损失和(或)费用
Application　　　　　　　　　申请
Ascertainment　　　　　　　　确认
Deferment　　　　　　　　　　延期,推迟
Details　　　　　　　　　　　细节
Payment　　　　　　　　　　　付款

Materials　　　　　　　　　　材料
Minutes of meeting　　　　　　会议纪要
Mistakes　　　　　　　　　　　错误

Named persons　　　　　　　　指定人员
Nominated subcontractor　　　　指定分包商

Objections　　　　　　　　　　反对
Obligations　　　　　　　　　义务
Opening up　　　　　　　　　剥露,打开

Operatives	施工人员
Oral instructions	口头指令
Partial possession	部分占有
Passes	通行证
Payment	付款
Performance bond	履约保函
Performance specified work	实施规定的工程
Person-in-charge	负责人
Possession	占有
Practical completion	实际完工
Price statement	价格报表
Priced activity schedule	已报价的分项工程表
Programme	工程进度计划
Provisional sum	暂定款
Quantity surveyor	工料测量师
Replacement	替代人选
Referral	争端事项
Restrictions	规定
Retention	保留金
Satisfaction	满意
Setting out	放线
Site manager	现场经理
Specification	技术规范,技术说明
Statutory requirements	法律法规要求
Sub-contractors	分包商
Adjudication	裁决
Architect's instructions	建筑师指令
Assignment	权益转让
Default	违约
Defects	缺陷
Delay	延误
Design	设计
Determination	决定
Direction	指令
Drawings	图纸
Extensions of time	工期顺延

Failure to complete	不能按期完工
Failure to conclude sub-contract	不能签订分包合同
Failure to reach agreement	不能达成分包协议
Final account	最终账单
Financial claim	费用索赔
Insurance	工程保险
Letter of intent	意向函
List	名单
Named	指定的人员
Nominated	指定的分包商
Objections	反对指定
Payment	支付分包商
Practical completion	实际完工
Sub-letting	分包
Substituted sub-contractors	替代的分包商
Successful project	成功的项目
Suspension of work	暂停施工
Warranty	保证
Withdrawal of offer	撤回投标
Withholding notice	不发通知
Sub-letting	分包
Substituted sub-contractor	替代的分包商
Successful project	成功的项目
Supplementary provisions	补充条款
Suppliers	供应商
Suspension	暂停
Tender	标书
Acceptance	接受标书
Amendment	修改标书
Confirmation	确认标书
Contractor unwilling	承包商不愿意
Contractor willing	承包商愿意
Date delayed	延误日
Documents	标书文件
Errors	标书错误
Extension of period	工期顺延期
Information	信息
List	名单

Qualifications	资质
Questions	疑问
Withdrawal	撤回标书
Tendering	投标
Termination	合同终止
Valuation	估价
Variations	变更
Verification of vouchers	发票证实
Warranty	担保
Withholding notice	不发通知
Written statement	书面说明